Planetary Service

Planetary Service

A Way Into The Third Millenium

Eugen Rosenstock-Huessy

argo books
jericho, vermont

Translated by Mark Huessy and Freya von Moltke

Wipf and Stock Publishers
199 W 8th Ave, Suite 3
Eugene, OR 97401

Planetary Service
A Way Into The Third
Millennium
By Rosenstock-Huessy, Eugen

ISBN 13: 978-1-62032-448-6
Publication date 1/28/2013
Previously published by Argo
Books, 1978

Distributed and managed by
Wipf and Stock Publishers.
For all orders and inquiries
please contact us at:

http://www.wipfandstock.com
info@wipfandstock.com
orders@wipfandstock.com

ecycled paper.

y Eugen Rosenstock-Huessy may be
s • RR 2, Box 366A • Jericho, Vermont
ilable books is found at the end of this

ooks in cooperation with The Norwich
erned with the future of such volunteer
ok. The Norwich Center was founded in
mp William James. Persons interested in
Center, Norwich, Vermont 05055.

True peace is neither the end nor the opposite of war.

— Teilhard de Chardin
Works, Vol. IV, p. 205

Contents

Preface

In an age much given to the vocabulary of individual and national *rights*, it must seem anachronistic to put forward the notion of specific duties - to be institutionalized in a specific service - to the world community. But rights can only be defended in the long run if corresponding obligations are willingly and freely assumed. Eugen Rosenstock-Huessy had the merit not only of articulating the philosophy of this viewpoint but of pioneering in its concrete application on two continents. In making available the first translation into English of the last of Rosenstock-Huessy's works, Argo Books has itself performed a service of no mean import to the future of planetary society.

As early as 1912, a bare two years after the appearance of William James's essay, "The Moral Equivalent of War," Rosenstock-Huessy published a plea that the German Army establish units for constructive civil work. A generation later, the sons of William James joined Professor Rosenstock-Huessy in Vermont to inaugurate a "new model" camp of the Civilian Conservation Corps, "Camp William James," whose history has been dramatically described in a recent Argo book by Professor Jack Preiss.

In 1978 it is clearer than ever before that the lives of people all over the world are, in Jay Forrester's felicitous phrase, "tightly coupled." The United Nations' Stockholm Conference established the extent of the threat of environmental degradation due to industrialization, population growth and the absence of effective corrective measures. Meanwhile, according to the International Peace Research Institute, the world is spending one million dollars *per minute* on arms and armaments.

Recognition is slowly dawning that there must be limits to individual, group and national egotisms, that a new ethos of cooperation and comprehension should be fostered by all, as a matter of urgent common interest. Such an ethos or outlook can express itself in many ways and on many levels: intellectual, spiritual, manual, artistic and so on. What is proposed in these pages is a kind of "basic training," open to young people all over the world, in intelligently directed, concrete service to the planetary environment.

A "Planetary Service," as the brilliant work service of the Weimar Republic amply demonstrated in the fateful interregnum between World Wars I and II, need not be monolithic, authoritarian or politicized in any manner. What would be required is a simple charter setting forth guidelines and purposes, and defining some simple procedure for support and accreditation. The "international civil service" which manages the agencies associated with the United Nations would be well qualified, under the able guidance of the Secretary-General, to carry out the assessments, administrative chores and support functions essential to bring about a true "Planetary Service" on a scale worthy of the problems and challenges to which the proposal has been addressed.

Public servants of impeccable standing have called our attention to serious *lacunae* in humanity's effort to preserve its essential resources of soil, water and air. To cite but one example: Klaus Lampe, Corneille Jest and Joseph Allen Stein, with the help of UNESCO's former Deputy Director-General, Dr. John E. Fobes, have documented the threatened de-forestation of mountain environments from Nepal to the Andes. Concerted re-forestation is a necessity for future generations. Cannot a worldwide volunteer force, with the official sanction of the world community, make at least a beginning on this matter? Is it not a proper "agenda item" for the United Nations General Assembly, for the Club of Rome and for international scientific associations?

From a narrow (but necessary!) legal viewpoint, it is not more difficult to devise a juridical framework that would take into account the domestic procedures and priorities of cooperating nations than it was for President Franklin D. Roosevelt to devise arrangements for the Civilian Conservation Corps that met the varying proclivities of the States of the United States of America. What is needed to launch the world enterprise is an aroused and focussed international public opinion. The publication, in the English language, of *Dienst auf dem Planeten* may help in a seminal way to bring about the necessary climate of opinion or, to phrase it in a manner which Eugen Rosenstock-Huessy might well have preferred, to arouse the necessary *passion* for worldwide service to the planetary environment.

Frank P. Davidson
May, 1978 Concord, Massachusetts

Translators' Introduction

Eugen Rosenstock-Huessy wrote this small book at the age of 76, after a career of scholarship spanning sixty years. It was his last work. He addresses it explicitly to the Germans who had asked him to write on the subject once more. So the original text is full of special references and allusions to German history and poetry. Rosenstock-Huessy was himself a German before he left that country in disgust in 1933. He became an American citizen, living, teaching and writing in the United States thereafter. He died here in 1973.

About this little book Rosenstock-Huessy wrote to a would-be American publisher in 1965: "I may have to rewrite "Planetary Service" for the English edition. The German text addresses the German public explicitly. . . . Hence an American translation might leave readers either cold or confused." He never did write an American version. His friends now feel there is both merit and urgency in a translation after all. "It is not really a good book, one of them said, but it is important." In translating we have left out some passages and sentences that had too German a context; they can all be found in the German edition.

Several chapters of the book were based on lectures, and throughout the book has kept the quality of the spoken word, or that of letters, as did much of the author's writing.

The readers may need patience; they may consider some passages obscure. In all his works Rosenstock-Huessy is out to awaken his readers' sense of time and timing, to make them aware of the quality of time, human time as opposed to physical time. He has always found it unnecessary to be understood completely at all times. And it is true: passages may become quite clear on rereading them at a different hour. The patience of the reader will be rewarded by the author's flashes of insight and his vision of generations of men and women working together throughout the eras of human history.

Mark Huessy
Freya von Moltke

Foreword: What Directs Me

This book has a catch. It will try to give its readers direction. It offers a particular path that leads through the areas in which we live. This offer breaks the rules of the academic world, in which the fields of research leave you without direction. Instead, they objectively project knowledge of the world and its parts onto your inner or outer retina. In the following pages we will be talking about this incidentally at most. Instead, I am presuming that the reader will take the opportunity to become aware of his own age-old destiny: "Truth has long since been discovered, has united hosts of noble spirits, that aged truth — seize it." (Goethe)

You cannot seize objectively. No one can be forced to do so. To be objective the reader may always escape into geography, natural history, philosophy, or statistics. He might think that he can observe the world from without with the help of these sciences. Here is a particularly idiotic example. There exists an expensive edition of a "World Cultural Series" printed in 1964 as part of a group of social science books. One volume is entitled "One World Divided." It presents us with the regions of world-cultures. There were eleven of them: 1. Europe, 2. the Soviet Union, 3. England/America, 4. South America, 5. North Africa together with Western Asia, 6. South Asia, 7. South East Asia, 8. East Asia, 9. Africa, 10. Australia together with New Zealand, 11. the Pacific Ocean. Now some people will have already found out that great confusion results when you use the words "culture" and "world." If you can get along without these words you will spare yourself many self-illusions. Our planet, the Earth, is part of an immense universe, and it harbors branches of the human race. And the branches of our race in, for example, Sydney, Australia, and Edinburgh, Scotland have much more in common than the people in Adrianople have with the people in Constantinople. The Lutherans in Sweden and the Lutherans in Calgary, Alberta, Canada, have much more in common than the Mohammedans in Northern Sudan have with the head-hunters in Southern Sudan. There are hordes of old ladies in Victoria, British Columbia, still living in the fashion of a Queen-widow as though this were the time of Queen Victoria in England.

So geography, culture and statistics are evasions because they don't tell us in what *hour* in the history of creation the various residents of our globe and the different readers of a book think they are acting, living and dying. The geographers are helpful to us in so far as they describe the Earth. But when applied to our own destiny these eleven areas are confusing, and no faith can be described in spatial pigeonholes like these. Whether I am wandering in the bush without the guidance of the sun and the stars, whether I am wandering on earth obeying the laws of the stars, whether I let myself be enticed further and further by a world beyond, as Milton's Adam and Eve did, or whether I am hoping to fulfill the law of our planet as a whole, these are four religions, each demanding very different sacrifices from their faithful. The religions confer different ways of thinking. And these ways of thinking have nothing to do with geography.

Every religion is characterized solely by its type of sacrifice. There are all sorts of human sacrifices, as well as sacrifices of animals, of money, of time, and of sanity. And the only religion which would be appropriate for all of mankind would require that we sacrifice a part of ourselves. This is therefore the only binding religion which has a future. A religion demanding human sacrifices à la Hitler, must disappear. In between, innocuous religions exist which sacrifice wax candles or the enjoyment of eating meat. The geographers and the cultural historians and the development-aid experts have eradicated the word "sacrifice" from their vocabularies. Therefore they know all about everything, and they know it very exactly. But they know nothing about the direction which we men have to choose. When they start talking about us they themselves modestly say that they want to remain objective, that is, to know nothing about anything important.

The result is that they are unable to know anything about us and have to talk nonsense about our future, for we certainly are never objects. What they know about us as objects concerns our hats, the soles of our shoes, our money and our wares and our diplomas. However they are daily completely surprised by our actions and sufferings. Julian Huxley of UNESCO was perhaps the most touching of these objectivists. The objective 19th century came to a very respectable close with him. As early as 1910, William James wrote his "The Moral Equivalent of War" to counter this mentality. In William James the spirits of the woods and of the bush, of the earth and of the heaven, of the planet and of the air spaces resisted the Darwinists, the geographers, the cultural statisticians and the world-historians. So my readers should not expect that objective scholarship will come spattering down upon them. We all write books. We write because, as we leave our mere individual life, we have to tell and carry the truth beyond our bodily passing into a common life. That is why I am directed by certain assumptions. I assume that my voice will attack false religions and, releasing the souls of my readers from them, will give the readers new direction.

Not only is there a right which permits the nations and the voters to have a voice in decisions, but long ago human speech itself was forcing this upon us as a duty. It is bestowed upon us by the destiny of the human race. The destiny of our race is not determined by science. The sciences are on the contrary subordinated to this future destiny. This future hangs over our birthdays, while the sciences are decorations for our old age. Without destiny there would only be females and males, yellow men and black men, Bostonians and Los Angelenos, eternally divided. But beyond these are the sons and daughters of the *One Man*. A common destiny for all children of man has been discovered for us. Ancient Israel wandered out of the Nile valley for all men; the new Israel has been crucified for them in the land of the Jordan. The calvaries of Bialystock, Auschwitz, Belsen, Dachau, Flossenbürg have tried to move us in Babylon's and Nineveh's false direction again, so I can not know what new heathen will laugh at this book. But even he is at least aware that our true destiny may dictate to the voices of science, just as it should command the shouting of politics or the sound of the arts. I have tried to subordinate myself to this dictate. This book does not look down on objects or things. It admits to the sin of remaining prepared here below to learn from above.

Four Wells, May 1965
Norwich, Vermont, USA

Eugen Rosenstock-Huessy

1. The Tone Which Is Missing

First the South German radio network and then my publishers have asked me to speak once more about what has seemed to me since 1912 to be the most important concern both in my own future and in the future of the peoples of the globe. I have been concerned with the formation of a common service into which every young person would be pulled in the course of life, so a new language could be spoken on earth. Its way of speaking would have to spring from a language common to all mankind, rather than from the national or class languages of past society. I wish to call what I am going to announce "The Tone Which Is Missing." If we are to have a future on this earth still — be it only a few hundred years — then it is urgent that this tone resound, lest fear and pride — those two powers which silence us — let the world go up in flames and freeze over.

I will have to order the story I want to tell according to what happened both to me and to everyone else in the course of the past 60 years. First I want to listen into the future to what is coming towards us, which could well be called negatively the barrier actively resisting something imminent. Let us call it the barrier against what is coming. I want to talk about the World Civil War, since this lost World Civil War is still with us, and about its lost battles which today are trying to prevent what is imminent. Then I want to tell you about the many beginnings which have been nipped in the bud. For a long time men of good will have been planting seeds showing us how we should plant the flower or the tree or the forest of the future. We must take note of the destruction which has already been wrought in the good will and energy of men. We have all witnessed not only the attempts to help the underdeveloped countries, but also the disappointments brought about by even these small efforts.

First I want to talk about the barrier against what is coming. To be absolutely safe, I am going to quote both the last Czar, Khrushchev, and a very old Pope. Both have expressed the same opinion in two different ways. The Church, ever since Jesus, has always honestly and openly said that only wars can change the borders between states. That is why in 1458 Pope Pius II wrote frankly in his "History of Frederick III": "States cannot alter their borders except by war." Several years ago Prime Minister Kekkonen of Finland met with Khrushchev and tried to ask him if little Karelia (the border province between Finland and Russia, lost in 1939) which was sorely missed by the Finns, might not be returned to Finland. Khrushchev who liked Kekkonen very much (Kekkonen visited him in the Crimea every year) smiled and answered: "Mr. Kekkonen, you ought to know from history that the borders between states can only be changed by means of war." People in the world today don't like to hear this, but I think that both the Renaissance Pope and the last Russian Czar are right.

And does preparing ourselves to live in a world without war mean that the borders between countries cannot be changed nowadays? For many people who dream differently this is a very unpleasant bit of news. It is in fact very alarming. Let's look at Israel and the Arabs. Tiny Israel depends on the hope that its borders — only 13½ miles from the Mediterranean to Jordan — will not be changed by war. If there is no war, they can hope that everything will remain as it is. On the other hand let us listen to the complaints of the two hundred thousand Magyars in the new state of Rumania, where you can be sentenced to twenty-five years in prison if you say that Hungarian is your native language. If these borders cannot be changed without war, then these several hundred thousand Magyars are condemned to downfall and destruction.

So it is not pleasant to live without the hope of being able to change borders. In fact, when I hear people always praising eternal peace, and when both the red Czar and the Pope agree that just changes in state boundaries can only be accomplished by means of war, then it looks very much as though it will be impossible ever to achieve justice between countries. How might it be possible to change borders without resorting to war after all? Or more cautiously, would it perhaps be possible in the future to change borders other than state borders, and how will that come to pass?

My teacher, Otto Gierke, the foremost European jurist of the 19th century, clearly saw mankind as a world society even then, causing people to say of him, half jokingly, half in astonishment: "Otto Gierke sees men without their skins on." Would it help us to look at the peoples of the world without their skins on — that is without borders — and how would we react if the peoples of the world could see themselves without their skins on? Well, for the last fifty years there has been an honest attempt to look at, to speak to, to address, to look in the eyes and to influence all men at the same time. I myself often speak on the radio which is a medium that theoretically can address the whole world. This is not without meaning for the fate of the world, and therefore the changes in radio broadcast declarations are very revealing.

The first broadcast "to all men" was sent in the Russian Revolution of 1917; the last broadcast that I know of, to all mankind irrespective of any borders, was sent on Maundy Thursday 1963 by Pope John XXIII. So both the atheist and the Christian are already speaking to men without regard for borders or skin, namely to all mankind. It took forty-six years from the start of the Reformation in Germany, from 1517 when the 95 theses were read by the few literate residents of Wittenberg, to the recording of the Heidelberg Catechism in 1563. It took forty-six years for the confession of the monk in Wittenberg to become the common property of many German princes and communities. The Council of Trent ended the same year, in 1563, and since then the religious denominations have become established.

Nowadays every movement wants to speak to all men; many people have taken that first broadcast as a model, and at some point the borders are being perforated. Even if Rumanians and Hungarians want to eradicate one another. Israelis and Arabs too will arrive at a point at which people will tell them "your borders don't exist." In spite of their horrible border in the middle of Jerusalem, Jews and Moslems have both been helping the same hard-pressed young governments in various parts of Africa, not cheek by jowl, of course, rather in "boundless" competition.

Here I would like to tell you a rather funny story about disappearing borders. A friend of mine in Chicago is an enthusiastic ham radio operator. One day he was talking with a ham in Tokyo, when the man in Tokyo said to him suddenly: "Please go over to your neighbor's house. He has just told me over his radio that he has locked himself in, and can't get out. Would you go help him?" And so it happened that the neighbor in Chicago was released, opened up, figured out, de-skinned, and unbound with the help of the man in Tokyo. That story is quite helpful; not much can be deduced from it however. It takes only a moment to let someone out of his house. That is probably easier than releasing and freeing someone living between the borders of other peoples from his fear and pride. First we will have to examine the purpose of borders a bit more closely, before we can summon the courage for a world without war.

2. A World Without War?

A world without war? Today you hear people recommending mere peace. As a result of the bomb, the word has gone around that although you can achieve your goals in a war, you will be eradicated yourself in the process. So war is being disavowed, and both war and the death penalty are being given up.

That should give us pause. Do you know what will be lost if war is abolished, and the serious values can no longer be measured by the amount of life a man is prepared to sacrifice for them? "And if you don't risk life, you never will win life" (Schiller). Is that not still true? Or another quotation: "Man grows stunted in peacetime, and lazy calms is the grave of courage" (Schiller). Is not this poet's song still true today? In 1910 the American Peace Society received a manuscript from William James (1842-1910) who was dying. It was read to the meeting for him. James spoke of the martial virtues which would have to be guaranteed if war should disappear. His address was to become famous under the title "The Moral Equivalent of War." In this speech there is an exclamation by this great and peace-inspired man which is still of particular concern to us at this moment: "Fie upon such a cattleyard of a planet" if war is to disappear without an equivalent. Men would sink to the level of tame cows and dogs.

The disappearance of war threatens us with the loss of the ability to distinguish between play and seriousness. Let us admit openly: war is the prime example of deadly earnestness, absolute earnestness. Any action in which I am prepared to risk my life resembles war. Even love sinks to the level of a game when there is no risk of life involved. As a cynic long ago said about love affairs:

"If it weren't for the consequences, it would be a parlour game." When love becomes a game, society dissolves. The individual who makes war on society, who becomes a martyr, or who convinces others, is also conducting a war as a soldier of Christ. Everyone knows that this kind of war will also change borders, just as the hermits in the desert changed the boundaries of countries. Since the crucifixion we have a better idea than before of what man can do.

On the other hand, anything which does not involve the risk of life resembles a game. Founding a club or holding a general meeting of a choir group is not the same as reporting for duty during mobilization. A mere club can be started painlessly and painlessly dissolved; there will be no tears when the club treasury is divided up among the members, or the choir group is disbanded in the face of an epidemic of laryngitis.

Today the boundary between seriousness and play is not usually seen as the boundary dividing war and peace, yet it is that same division. In a game I can say "one try does not commit me," but in time of war anything I do happens once and for all. Whether we keep them apart by calling them war and peace does not matter to me but we must keep seriousness separate from play.

We must be aware of the fact that borders between peoples, borders between religions, borders between men, borders between sexes, or borders between generations, are serious only if they are final, if they can be changed only at the risk of life. Otherwise it is a game. Nothing is truly serious which can be changed in the course of a phone call, like an invitation to a party. But the Easter holiday cannot be shifted or cancelled. It is either more "final" than we ourselves are, or it has been abolished. History is made up of events which are final.

Of course it is true that most of the people whom I meet studiously obscure the difference between seriousness and play. They try to treat war as a game. Like my contemporaries in 1914, they want to say playfully, "Honored Sirs, declarations of war are being accepted at the window."

But games can also be horrible. For example, in 1933 I led 65 American mayors through Berlin. After I had accompanied them to the new Nazi potentates in the Tiergartenstrasse, talking about these criminals, they said to me in shocked surprise: "Why they're playing cowboys and Indians!" These were sober travelers who had just returned from Russia, which was terribly earnest, and they could sense the difference. I was reminded of James Fenimore Cooper myself when I last negotiated with the newly-baked Nazi bigwigs. I wrote then that the apparition of the "thousand year Reich" would last for twelve years, because as of 1945 the people born in the baby-rich years after the First World War, 1920-25, would be able to make themselves felt. The very slogan "thousand-year Reich" reveals the Fenimore Cooper character of the Nazi torture. When reputable government attorneys went around personally setting fire to synagogues, the naughty little boy was emerging from inside stuffy officials. Only this additional facet entirely explains the Nazi horrors. They were so much like an apparition, that a professor in Breslau, Eugen Kühnemann, called out: "That is just how fast the Third Reich will pass!," as he noticed a troop of SA (*Sturmabteilung*) men marching past the window of his lecture hall.

Most people pass over this obvious truth these days, but anyone who wants to understand the disappearance of the nation-states in the course of the two World Wars has to take notice of the false-bottomed aspect of the Nazi chimera. Hitler's suicide and his rampage against the five thousand finest men and women of the German Resistance after the Allied landing in Normandy can only be understood as theatre. Hastily dragging your political enemies into your own grave no longer has anything to do with politics. Those closest to Hitler knew perfectly well in October 1944 that the war was lost, yet that was when they really began to execute and murder. This is very important in our attempt to distinguish seriousness from play. Here we have a group which worships war but is on the verge of defeat. And what do they do but deliberately murder all those who will be most needed after the defeat occurs. The reader cannot find a better example for how these brains, single-mindedly bent on war — a war between nations — had arrived at their ultimate spiritual boundary. The decision to make war had defeated itself. For Germany the ancient game is over. It is over because the nationalists falsified war and revolution into an open game, a game without end. Thus the ancient rule that peace could always be concluded by a new government no longer applied. Neither war nor revolution will ever be "played" in Germany again. Chess champions take chess very seriously. Nevertheless chess is a game. Goebbels never took war seriously, yet war remains as serious as love.

Both love and war are perilous; they produce corpses and invalids. In both love and war we have to risk our lives. New life always requires the risk of old life: the death of soldiers for the life of nations, the death of mothers in childbirth. Our lifeblood nourishes new life. The words "old" (*alt*) and "parents" (*Eltern*) in German share their roots with "alumnus" and "alere" (*nourish*). The old nourish the young, and that is what makes them "old."

But when they tried to make a thousand-year Reich, a non-aging empire, out of war, they descended to the level of a devilish game. A millenium is useless to mortals like ourselves when we stare at it as a whole. "Our life is encompassed by a narrow circle, and many generations arrange themselves in an unending chain of their existence" (Goethe). The big game of cowboys and Indians seems to have brought about only a millenium of defeat.

But the two World Wars did not get named after any particular country. They were world wars. The fruit of these world wars is the unity of the globe. The division of Germany and Korea has pulled the world together like Guericke's spheres. Divided Germany and divided Korea are pressing the world together just as the two half-spheres of Magdeburg's mayor were held tightly together when the air was sucked from them. The meaning of these events cannot be found in Germany or in Korea anymore; their division demonstrates that the globe is one.

So we stand in the middle of a puzzling time. Individual states have ceased to be the main actors on the stage of the world. Hitler's attempt especially has made it clear. But how can we stand to live in a world in which nothing is absolutely serious, and in which, for that reason, nothing new can be born anymore. Must all of the current borders petrify, as only war could change them? Must men like the Nobel prize winner Luthuli in South Africa perish just as another Nobel prize winner Ossietzki did under Hitler? Then we peaceful innovators will have to remain locked inside the borders of South

Africa or Switzerland, and only because we cannot wage war anymore. This noble man Luthuli was appointed Honorary President of St. Andrews University in Scotland, and cannot even receive his mail from there. Must a Nobel prize winner be condemned without cause, just because he is confined within the national borders of South Africa which are eternally secure in the absence of war?

Without forces to rid us of our boundaries in the future, I would find the condition of simple war-lessness hateful. I utterly disagree with the protesting atomic scientists' claim that we could remain vital if we simply abolish war. There will always have to be some way to release those forces which have been changing borders since the time of King David and Agamemnon. I yearn for a condition in which I would not have to suffer from the nightmare that all current borders might remain in existence until doomsday.

But they would in fact be unchangeable, were we unable to create border-crossing points without resorting to war. The immense dilemma facing us today is not a lack of insight that the bomb cannot be thrown. No one is making that mistake, neither the Pope nor Khrushchev. What is missing is a powerful and enheartening means of changing borders without war. We are going to have to overcome borders without the bloodshed we have been used to in war. Perhaps it would be helpful to remember that we mortals have always been hemmed in by two kinds of prisons, the first being the world. That shows up on the map. There are houses and gardens and fences, the boundaries between towns, borders between countries, and finally even the borders between continents. These are all borders between the spaces in which we live. Once when I was an eleven year old boy, I drilled a hole in the door of my sister's room, and was severely punished for destroying the lovely door. I learned how dangerous it can be to move boundaries.

Luckily there are other boundaries. In English the word "neighbor" does not mean just people living on the same street, but also the person whom a living man needs most at a certain hour of his life. (German has two different words.) There are also borders in time. The Nazis built borders in time as high as borders in space.

Authors were required to put the year of their birth on the title page, just as if they were part of a stud farm. This allowed any stupid little boy to say, the author is too old for me, or any stupid old man to say, he is too young. Thus a border was created.

In reading Shakespeare we can see he felt that the misfortunes in his tragedies could not be solved in a single generation. In fact all of his last plays are written around this fact. Shakespeare offers the solution that, while it was indeed impossible for a single generation to find a way out, the second or third might do so. "The Tempest" is a good example of this in which Miranda and Ferdinand live happily, although the older generation is at daggers' points. It is the same in "Cymbeline," as well as in "A Winter's Tale," and this also is the promise in "Hamlet." The underlying thrust of "Hamlet" is just this: he sets right the times which are out of joint by risking his life, in order that life can continue after his death. So borders can be overcome in three generations.

These borders that run through time rather than space will have to be discovered anew. And most of the modern poets have simply passed over Shakespeare's most serious questions, those about cooperation between generations over time. Modern man and his philosophy began with "I," that is with the young student, the wandering youth, or the naive individual. They recognize only the bodily phantoms of heredity, or maybe a father's stern treatment of his son. The individual is supposed to function without a before or an after, without ancestors or descendants. Today people write about an itch or an upset stomach lasting no more than a single day or perhaps a mere week. In addition the modern public feels that the story of the fulfillment or destruction of an enterprise carried out over a series of three or four generations is only an arbitrary grouping of unrelated events.

Goethe once said: "We are barred from looking beyond." This is also true of our view of other times. Looking backward in time we overlook the promise that called us into being; looking forward we cannot see the outcome. Five year plans are considered to be something new and exciting in 1965 on a continent where the forests of Ceasar's Germania have maintained their 25% share of the landmass since 1100, and on which the Altamiran cave paintings look at us after the passage of 3,000 years just as they have always done, since a time before there were either Romans or Germans.

The death of literature nowadays probably has something to do with its shortness of breath. The so-called "Lives of Jesus" offer a funny example of this. Up to 1800 Jesus was regarded as the fulfillment of all expectations in human history, the Promised One foretold by the prophets. He had been expected as long as man had been alive. Then the 19th century came and robbed him even of the generation which had preceded him, namely of John the Baptist who announced him. Similarly the 19th century claimed that the generation following him had thwarted or frustrated Jesus' true intentions: Paul was supposed to have ruined beautiful Christianity. The 19th century made the coming of the second Adam into a bit of "contemporary existence" — for you certainly could not call such a mid-summer night's dream "life." Was the man for whom the nations had waited so long merely a contemporary of mad caesars?

The true man reaches across the times from the dawn of creation to the furthest future. And only that man in whose name such streams of power are united is of concern to us because borders fall before him.

Perhaps those who occupy themselves with prehistoric times can help. After all they talk freely about millions of years. This makes it easy for us to forget the physical borders which we see on our maps which suggest to us that almost every tiny island in the Adriatic preserves its own individual fauna. But do unending timespans really make the horrible future of frozen borders any more bearable? Again I would remind you of the words of both the Pope and the red Czar: only wars can change borders. The currently reigning Napoleon IV of France (de Gaulle) agrees.

The question is a serious one. Both Louis XIV and Frederick the Great put the words "ultima ratio regum" (the last resort of kings) on their guns. As a Prussian artillerist I believed that these inscriptions were a solely Prussian stroke of genius. The horizon with which I grew up was still that narrow. Nowadays, however, no one can harbor the illusion that only their own community will throw bombs. And consequently well-meaning people have been crying "Stop war!" Quite a contrast to old Bodelschwingh of Bethel (a German village for disabled people) who taught: wars will always recur.

He would take care not to say that today. But you cannot get rid of an antiquated "yes" simply by substituting a straight "no." Saying no to war would only make way for a new "yes." There is an iron law "if we do nothing but flee we cannot escape." Our age, so keen on natural laws, ignores this iron law, thus making war unavoidable. It will be possible to abolish war only by creating some other way of altering borders.

There is a particularly pompous example of the arrogant presumption of the simple abolitionist approach. The former president of the University of Chicago, Robert Hutchins, and a staff of selected colleagues have settled in Santa Barbara, California, for the purpose of studying the political future of the world. One of his colleagues is Walter Millis, a respected military author, who has written books both on the second World War and on military affairs. He has published a study entitled "How can we conduct the Peace?" He mentions seven authorities out of the large number of advisors he interviewed for his study called "The Peace Game" (*Saturday R eview,* September 24, 1960). This study represents the most serious attempt to organize eternal peace which has been undertaken since Immanuel Kant. The study starts: "Let us begin by assuming that war has actually been abolished." Then he attempts to think beyond the hydrogen bomb. Unfortunately all he does is think. You cannot learn anything by just thinking. The bomb's saying "no" to war does not mean that the most important point has been considered, the one which allows us to publish studies about eternal peace these days, the point which even Stone-age man considered and the starting point of this chapter: borders. Nowhere does Millis ask how one could cross closed borders without resorting to war. Ever since prehistoric times, the "hostis," the foreigner was both enemy and guest. If he came unarmed and defenseless he could pass any border whether it existed around a camp fire, a wood, or an oasis. If he appeared armed, he was resisted. We are looking for this sort of "switchable" border again ourselves. For even after war has dried up, we will still have to distinguish between guest and hostis, friend and foe.

This entire book will be searching for that which causes borders to be switchable. One thing is certain: we have no hope for abolishing war until we accept the framework of a universal, planetary method of crossing borders between all peoples and all countries. The old division between peaceful association and aggressive association must be overtaken by a new division. From Millis and also from pacifists who are plagued by guilty consciences, all you hear is an immense, global "no" to war. But what is going to happen to the borders then?

The future paths towards peace, like all new paths in the beginning, are probably very insignificant-looking. This is how God has ordained it for all fruitful seeds of human thought. The "no" to world war resounds as loud and immense as the World War istelf, but the effective "yes" to a planetary peace will probably be insignificant and quiet. This is appropriate for an intimidated humanity reappearing from the bomb shelters of the last war. Teilhard de Chardin objected to the big planners of all times, pointing out how only that could become great and generally acknowledged which was first seen by no one. He put it this way: "No one recognized the first Roman as such." We, however, who after all have been moving along the path towards planetary peace as slowly as a tiny snail for over fifty years, may perhaps be allowed to tell how we have already climbed the modest seven hills of this future Rome in the meantime. It will not be anything big like the "no"-dream, or the eagerness to turn everything upside-down which comes from the brainy popinjays who even openly presume to call their empty speculations about peace a game, the "Peace Game."

We may be wrong about our service corps but the blows which we have received in their pursuit justify us in regarding ourselves as legitimate children of an emergency. After all, I have spoiled my career in the cause of peace service corps several times in both Germany and in the United States. That is of course the least one could ask of a person who is plotting to stop three thousand years of war from having heirs.

A type of magic trick exists, which someone might use to

secure eternal peace instantly. The trick lies in multiplying the number of passports we own. Nowadays every person has a right to a passport from his own country. (Only Swiss ladies who have married foreigners can enjoy two passports.) If we were all issued all one hundred and forty passports of the states in the United Nations, then we would not be able to wage war, because we would all be citizens of each of the warring countries. Although this sounds like a joke, it is not a bad one. I have personally helped arrange a case in which a high judge already possesses the passports of two great powers. Let us expand the diplomatic corps infinitely. Project the situation in Jerusalem, where people with diplomatic passports can travel freely between Zion and Jordan, multiply the situation by ten thousand, and then when the majority of us were diplomats all borders would break down or else seem outmoded as tails on a formal coat. Every reader should study for himself the difference between the supposed abolition of war by means of peace fanfares, and its eradication through the fact that no one would belong solely to any one of the warring parties. Can clerics and marxists build a coalition? In theory, no, but in Austria they have done it and thus have abolished civil war.

One might not have to touch the borders themselves much, no matter how crazily they may be drawn. One only need find ways of superseding them. Perhaps the hottest questions facing working, courting and testifying mankind today are: "Can I emigrate? Can I marry any child of man? Can I work anywhere? Can I print my book in Germany when I cannot print it in Russia?" These have been very pointed questions for the Soviet Union in the last twenty years.

But they are not just pointed for Russia; they have been just as pointed for me. A large proportion of the German intelligentsia is currently living abroad, but they are part of the realm of German language just the same. It became unbearable in 1933 when Germany withdrew into herself and regarded the rest of the world either as a battlefield or as enemy territory. For that reason the Second World War broke out in 1939. It was unbearable that only the words spoken inside Germany were considered German. Goethe's and Schiller's German had to go somewhere else, and as you know it was carried into many parts of the Earth by this emigration. There, people continued to speak German even after the fount of the language in Germany had dried up. And those of us who have been writing and speaking to the mother country since 1945 played our part in starting the German language living again, in letting it renew itself. After 1945 the effective and fruitful German language was reborn from abroad. It is perhaps too early to talk about this wonderful "feedback" of the worldwide realm of the German spirit to the mother country. Considering the specific case of the German Lutherans, saved from starvation by the American Lutherans, even the dogged German Lutherans from Neudettelsau who continue to hate the Jews, may recognize that their Lutheranism could only be reconstructed thanks to the German-Americans.

They should truthfully answer the question: "Did Hitler speak German?" If what he spoke really was German then the Lutherans have been impotent for the past four hundred years, and they have not succeeded in making the Lutheran Bible the mother tongue of the Germans (which actually did happen). But if it was not German that Hitler spoke, then this precious language has apparently been renewed by the people throughout the world who remained faithful to it, and this renewed language has been able to clean the Augean stables of Hitler's German. *Tertium non datur* (there is no third possibility).

Languages must go out into the world. They are not just vehicles for presenting dinners on an inn menu, but rather they carry out into the world the Word through which we can become children of the same father. And this is why they are spoken, even if in the eyes of minors they seem to exist just to cover menus. Therefore the renewal of speech in the mother country by the German spirit in foreign lands represents a great ascendance over borders, and has proven an asset in the struggle to regain German national sovereignty. Something as delicate and fragile as language cannot be locked up inside borders.

The poor Albanians are probably today the last people in Europe to have to practice encapsulment à la Hitler. Of course the Albanians with their isolated language have always been an unusual fragment. The proud latin word "imperator" continues to live in Albania as the toneless word "mbret." Albania became an independent state in 1912, and the Prince of Wied became its "Mbret." Actually more Albanians live in Southern Italy or in Massachusetts than in Albania. And what has happened to these poor Albanians since 1954 is even worse: they have to learn Chinese for political reasons: they have to listen to it and to read it. All this because they have turned away from West Rome and East Rome, opposing both Washington and Moscow. So they have to reforest the truncated remains of the Albanian language with the jargon of Chairman Mao from Peking.

That is an enlightening story! No language can remain shut within its spatial boundaries. So tiny en-chinesed Albania is a great example of the "horror vacui" (fear of the empty), the inescapable yearning of all languages to get to each other. If the reader is becoming impatient and thinks that is old hat, he is wrong. The boundaries of a language are also the borders of peace. For example, my books are still not allowed behind the boundary of the Iron Curtain. There do exist limits to the power of language.

But more important is the fact, seldom considered these days, that up to 1840 family members had no way of communicating across national borders, and that this is no longer true today. A decisive event has occurred in the meantime which has done more to further eternal peace than the bomb. There is a treasonable institution which functions in the full light of day, like every well kept secret; everyone knows about it, and no one thinks about its amazing effects: the postal system and its rates. I will have to digress, and say more because technological mankind has already forgotten about it.

When I look back, I am always overcome by the situation of mankind at the time of the Council of Nicaea. This was the Council in 325 which established the Trinity, and became the first ecumenical Council of the Church. No other Council has been so highly regarded. The meeting place was just opposite from Constantinople. But in spite of its fame thirty years later, in 355, the Bishop of Poitiers in Gaul had not yet heard about the decisions of this Council.

At that time even the words of the Church penetrated the Christian world to such a limited extent. Nobody thinks about that any more, and therefore we forget that things did not look much different in 1830 than they did in 350. An Irish dockworker in Glasgow, Scotland and his mother in Cook, Ireland could not write to each other as long as they lived. The cost of sending the letter would have equaled the mother's monthly income.

We can find the story of our Babylonian confusion of tongues in the Old Testament. There it is told that it began with the building of the tower of Babel. However in the New Covenant, the separation of languages caused by spatial distance is supposed to come to an end. The insurmountable barriers which have been built up by our inertia will melt away. Only kings could regularly exchange letters in earlier times. And it was not until Roland Hill established a universal postal rate in 1840, so that a letter from Ireland to London cost no more than a letter sent within London, that a mother in Ireland could correspond with a son in Scotland. That was the point when the disintegration of borders began, which is bemoaned nowadays by ministers of war. So it is not true that the bomb was the first or only thing that made wars impossible. Their abolition has been under way for a long time.

So the atomic physicists' debate over peace sounds hollow to me. The achievement of uniform postal rates (followed by telephones, radio and television) seems much less anemic. Since 1840 our whole inner man has been increasingly freed from the coincidence of a particular location. The new world of Roland Hill's postal rates sprang from the bold denial that a letter from Boston to Cambridge must cost less than a letter from Boston to San Francisco. Contemporaries of Hermann Melville considered this a crying injustice and an inexcusable mistreatment of Cambridge. The Bible tells a similar incomprehensible story about wages in a vineyard. According to the story, an hour's work was to bring the same wage as a day's work. The workers who were putting in long hours cried: "Scandalous!" The people in Cambridge raised the same cry in 1841: "It's scandalous that the people in San Francisco don't have to pay more than we do." But one of the opponents of the uniform postal rate clearly expressed the result of Jesus' revolution in the area of wages and in the life of the emerging world postal union. The reader will be astonished by the statement of this enemy of modernization. Such clairvoyance at the first moment of a great change is often breathtaking.

But before I illustrate our first moment in 1840 with his clairvoyant comment, I would like to call your attention to a famous example which reflects the historical process of the great revolutions. Beaumarchais' "The Marriage of Figaro" was read aloud to the King and Queen of France eight years before the storming of the Bastille, July 14, 1789. The King responded indignantly: "If this play were allowed to be performed, you would have to pull down the Bastille." The play was performed in 1784; the author was imprisoned in March 1785; in August the King and Queen received him as an honored guest; in 1789 the Bastille was stormed. In 1793 the King was guillotined.

And what was the comment in 1840? The perspicacious opponent, recognizing the revolutionary character of Hill's uniform postal rate, cried: "That would mean the abolition of time and space!" And this is exactly what the bomb makes obvious. "World" history has ceased to exist on earth. Because in "world" history a deep gap lay between the area where your home was and the world. You could go out into the world; you could conquer the world, but your homeland had a special claim on you. We can no longer conquer the world because we do not have any starting point that we can call our own, a point the "world" cannot reach. Our home can be flown over, can be seen from above; the world postal union has made all times and spaces equal; hence we now live on a planet, and no longer in "the world." On the planet the differences in time and space which have hitherto made our home different from the "world" have been written off to the point of nonexistence, and so must be entered into the balance sheet of our lives with a zero. Every industrialist knows how true this is for his products. I recently bought a piece of clothing in rural America; it had been woven in Bielefeld, Germany, sewn in the Phillipines, and sold in Boston. The one thing which has not occured to the businessmen and physicists however is that there is no point in continuing to talk about a "world" if all distances are equal to zero.

Albanians learning Chinese is an excellent example of the lengths to which even the most cut-off group will go, to resist being isolated within its own borders. And this should make it apparent to us that there must be ways to sound that tone which is missing, that tone which can overcome borders and sound across them, the tone which also makes it possible for me to seek my wife at the antipodes, and to help a neighbor in Chicago. Let us open our eyes! Are the national borders our own borders? People living next to one another in large cities will probably have the most difficult border to overcome. They look right through one another. Of what concern to me are those millions in my big city? In spite of them I am going to show that, given time, even the impossible can become possible, even the biggest stumbling block will disappear.

But to end my present train of thought, let me say something to do honor unto the borders. The borders enshrine everything with which its Creator has entrusted mankind to date. They reveal all of the causes for which men have risked their lives. National borders are not as contemptible as we like to make them nowadays. Although all borders — even the borders of private property which for instance we have decided to defend against communism — are honorable only as long as men who believe in them bear sacrifices for them. The word "boundary" is colorless, so to speak, if you meet it in philosophy or in jurisprudence, or in politics, but for a colloquial meaning we must remember Goethe's "Limits of Mankind": "Our life is 'bounded' by a narrow circle." We mortals feel the greatest reverence for things revealed to us so far, and all things revealed hitherto have led to the construction of borders. Therefore, at this moment I want to warn myself against presenting borders only negatively. No, they are venerable insofar as they allow us to hand on things entrusted to us. You can receive your heritage only because borders exist. The orientation which mankind needs would remain undetermined without them.

Of course the border cannot reveal whether what is being handed on is still alive or not. But we must faithfully remember that at one time something alive had received borders, like a body its skin, or a man his shape. Borders may only be given up when it becomes essential. Razing borders becomes essential only when something calls, as important and valuable as that which previously established the existing borders. The most glorious expression of the yearning to outbid, even to surpass all borders, may have come from the great Frenchman, Charles Péguy. He was born in 1873 and was forty-one years old when the first World War broke out, in which he fell during the battle of the Marne on September 4, 1914. He was actually too old to be a soldier, and his is a remarkable case: here was a man who had dedicated his life to the struggle against old borders, and who then fell in defense of the old borders of Europe, as did all the soldiers killed in the war of 1914.

This contradiction lies across my entire generation: we were all soldiers for the borders of our countries, and we were all already burned by the insight that many borders would have to fall. And these are the words about borders which Péguy left us before his death: "And must I, to rescue from the eternal flames the bodies of the damned who despair in torment, give up my own body unto the eternal fire, then, God, throw me into these eternal flames: and need I, to save from the torment those souls damned to stay forever distant, and who despair in their estrangement from You, let my own soul stay estranged, then let my own soul move into the eternal distance, for we can only save our souls together." Here you have one of God's heroes who will not leave any borders standing, not even the borders of hell. And everyone should first pause, considering Péguy's battle, and this contradiction. Only then shall we become serious.

Let us keep in mind: Poles and Germans and Russians and Dutch and Britons and Cypriots are obviously divided. But it is true nonetheless that "we must save ourselves together." We must reach God together.

Because the technical insight "there can be no more wars" is being preached to us, we must give our souls practice in surmounting borders. Technology cannot achieve this, for the shadows and ghosts of the past are powerful. Anyone who considers them vanquished by technology will end up like the young Frenchmen in 1929. They published a pamphlet at the time, "*La guerre ce sont nos peres*" (The war is our fathers' business). They published it just when the depression so overshadowed Europe that the Second

World war began to follow the first. These sons who had renounced their fathers and freed themselves from them in 1929, became tangled anew in the faults of their forebears just when their pamphlet derided them. Not even our times can escape the fourth commandment. Shakespeare knew that it takes several generations to acquit men of guilt. The European peoples — the Americans and Russians too, of course — repeated this pattern in the period between the two World Wars. They wanted to forget the war, and that is precisely what they should not have been allowed to do. For as they only turned away from it, it returned, larger and more furious. Those few who knew in 1918 or even in 1914 what kind of war it was that dragged the Europeans together and rushed them to destruction, these few have shown us the way out today. Today we must start from the cold war, and not from an apparent peace which really means only that no bombs are being thrown.

Marshal Lyautey learned of the declaration of war in 1914 in Morocco. This wise old man said at that time: "So civil war has broken out." People stared at him astonished, and the patriotic Frenchmen in his retinue found this very puzzling: civil war? Lyautey responded quietly: "A war between Europeans can only be a civil war." Baron von Schön, the German Ambassador in Paris, wrote on his visiting card, as he took leave from Paris: "this is the suicide of Europe." And Sir Edward Grey, the English Foreign Secretary, said at that time: "The lights are going out over Europe."

At the end of July, 1914, Nathan Soederblom, future Archbishop of Upsala, traveled to a university convention at Kösen in the same train compartment with me. He was beside himself because the German universities were agitating for war instead of working to keep the peace. He felt, as a Swede, that he should represent the spirit of the Christian university as well in Paris as in Leipzig: he was a professor of both universities, and he believed that teachers, particularly theologians, were there to fight to the last for the unity and peace of this new brotherhood of the world. He was most dissatisfied with the Germans, and I fortunately have not forgotten his dissatisfaction for the rest of my life. The train ride marked an epoch in my life.

But an occurrence in Heidelberg in 1908 will serve as a good example of how naively people then regarded "war" as war between stranger and stranger, between people of other faith and people who were different. At that time the "Zeppelin" had crashed in Echterdingen and burned up in the meadows. Within a few days the whole German populace donated enough money to build another Zeppelin. The enthusiasm and participation was enormous. But the boatman on the Stiftsmühlen ferry above Heidelberg knew it even better, and told me as he ferried me across: "Of course the Zeppelin was never really burned up — we're only telling the French that, so that they'll think that we don't have a Zeppelin any more." That was the extent to which everything was subordinated to the polarization between nation and world.

Marshall Lyautey knew that a war between Europeans was a civil war, but the civilian patriots and the German industrialists derided this insight. Yet even nowadays it is of extraordinary practical importance that Germans take these words "world civil war" very seriously. I think that the words "civil war" are hated in Germany. Actually they are only a translation of the word "revolution." Both war and revolution would benefit, were we to recognize that they are both forms of civil war. Today, after the reconciliation between Germany and France, it should not be hard to grasp that there should never again be a war between France and Germany, as we realized already in 1918 after we stood in front of Verdun for two years. More than 800,000 men had fallen at Verdun, 350,000 Germans and 450,000 Frenchmen. With that, any future between Germans and Frenchmen had become meaningless.

Such an experience has the advantage that people on both sides of the border reveal their characters as fellow citizens, and the name "Frenchman" or "German" changes from a noun into a simple adjective, "French" or "German." This is a philological truth long practiced in America, but unfortunately unrecognized in Europe. The European philologists know too much Indoeuropean to acknowledge that the German language will necessarily yield in the future. I am of Jewish origin, for instance, and proud of it, but I refuse to be considered a Jew. The word "Jewish" applies in my case: the expression "a Jew" does not. Just as often the adjective "Polish" may be correct, while the noun "Pole" is incorrect.

Whenever a border falls — that is what I want to tell you — a noun becomes a simple adjective. There are German-speaking Americans, and Spanish-speaking Americans — they are both Americans, though, and the difference between them may only be expressed by an adjective. Virgil's XIIth book of the Aeneid demands the same for the Romans. It is a great law, and there are Europeans who speak German and who speak French (and who usually live in or come from Alsace-Lorraine). So we should no longer describe the borders with nouns. As soon as we use an adjective instead of a noun for people on one side of a border as well as for those on the other, we suddenly realize that the border is not absolute. It has become relative, and we are going to have to make use of these relative borders in the case of the two wars, as otherwise they would be insurmountable.

But that is not enough. When Lyautey said that a war between Europeans could only be a civil war, or when Baron von Schön correctly wrote on his card "This is the suicide of Europe" (the nationalists and the Nazis made his life impossible later for this simple, courageous truth) they were indicating something more than just the world civil war. Namely, if men who are our brothers are living on the one side of the borders, then the world has ceased to be world.

To me the greatest event of the two World Wars seems to be the creation of the condition in which it is no longer permissible to call the globe simply "the world." What should one call it then? In 1946 in London (that is outside the areas that I have lived in or live in) I published a pamphlet. It was requested of me, and was called "Planetary Man." Its subject is man living on one planet. And if the men of this world have learned mutually to accept each other in a world civil war, in a world of revolutions, then this world has ceased to be merely "the world outside," and we may begin to claim the good planet Earth as our common home.

3. Planet - World - Earth

When a child learns to pray it sets the living God into heaven and asks that his will be done on earth. When a man learns to act his goals are set into the world, a busy world harboring an indivisible mixture of good and ill, which for him consists of the outer world and the inner world. So the "world" refuses to be divided into God's heaven and our earth. The astronomical celestial world is just as ungodlike as the terrestrial world. Or the terrestrial world is as godlike as the Milky Way. The word "world" is absolutely incomprehensible for theology, just as "on earth" from the Lord's prayer is only comprehensible when the earth is directed by heavenly will. Of course on the path of man's salvation, the words "world" on the one hand, and "heaven and earth" on the other meet one another at such close quarters that they cannot get out of each other's way. On pain of the end of the world, on pain of the collapse of the heavens, on pain of physical destruction we have to get the words heaven, earth, inner world and outer world to relate to each other properly.

Good genii ever since Copernicus have been preparing a way out of the collision of "heaven and earth" with the "world." We men live neither as lords of the earth, nor as servants of heaven. We do not live as children of the world either, as bits of cosmic dust. No, we live on a planet. What does that mean? Well, it is just beginning to mean something. So far the heavenly minded and the worldly minded are contesting our right to do so. Actually our planet is neither distant world nor native earth. It is a celestial body that moves all of us together with our antipodes around the sun. It does not confront us as a strange world: we are part of the whole.

As yet we are dependent on the coincidence of a location here or there. But already we receive our orders from the planet as a whole. While our whole species has lacked affinity so far, we have now been given a warning, since war cannot be waged anymore: "You are already one and united. This little globe is all that is left to divy up, compartmentalize, and hand out. Heaven comes to your earth whenever you recognize it as a planet, as the heavenly dwelling place of all the sons and daughters of Adam." To the world your home country and property stood in opposition. On the planet, however, we are all equally at home - left and right, America and Russia, are all equally far from being able to presume to be our exclusive home. *The planet is our home.*

Since that is difficult to understand I will tell you a true story about the outbreak of war after Pearl Harbor in 1941. Hitler declared war on the United States on the 8th of December, because the Japanese had attacked Pearl Harbor on December 7th. Pearl Harbor is in Hawaii, Hawaii is in the Pacific Ocean, fairly far south of Tokyo and Japan. Forthwith an arctic explorer, the Greenland expert, Vilhjalmur Steffanson, an Icelandic American, went to Washington to the War Department and said: "Dear people, you have just nearly lost your fleet at Pearl Harbor. Do you know why this has happened to you? Because you didn't believe that the earth is round, that it is a planet." "Oh, said the people," we learned that in second grade." "Well, you may have learned it in school, but you didn't believe it, or you wouldn't maintain that the shortest route to Japan went through Hawaii. Actually, because the earth is round, the shortest route to Tokyo runs via Alaska and the Kuriles. You not only have to know that the earth is round, you also have to believe it." That is how it is with the world and the planet. We have all known that the earth is a planet, but only after the Sputniks circled the earth in 89 minutes did we also have to believe that the earth is a planet.

That has enormous consequences for Germany. For if the earth has actually become a planet, then no one is in the middle any more - either in the middle of Europe or in the middle of the world. We are all on the periphery, for the planet itself is moving, and we are continually entering a changing relationship on the surface of the planet. The German nation has always been protected by hegemonial powers in the last two hundred years. Prussia and Austria held their protecting sword over the nucleus of the German people. That means that the German powers which existed on German territory were by no means only German. They were a mixture. Both Germans and non-Germans were necessary to keep Germans alive as Germans in the middle of the continent. But in 1913 the one hundredth anniversary of the Battle of Leipzig was celebrated as if it had occurred after Bismarck's time. The Austro-Hungarian Crown Prince left the celebration indignantly saying: "After all the Germans were the ones who fought under Napoleon at Leipzig, and it was the Austrians, Hungarians, Croats, Prussians, Poles, and the Russians who were victorious over him."

Well, what has happened now? Russia has taken over Prussia's place, and America Austria's, and Germany exists today only on the embrace of two world powers. Washington and Moscow have replaced Berlin and Vienna. And so it is extraordinarily important to persuade these two world powers to consider themselves planetary powers instead of world powers, for then they all have something very important to do together: namely to keep the peace. If they are mere world powers, then world imperialism will just continue. World politics means that they want to snatch booty away from one another like wild sharks. If they knew that they were living on a planet, however, one might hope they would realize that they have to make arrangements. Everyone realizes this, once he stops bawling the same old political songs.

When you can fly around the world in 89 minutes, it does not make much sense to blow up bridges over the Rhine. Transforming the world into a planet would certainly relieve the German situation,for the subjected, conquered and divided people could appeal in the name of the planet against Russia's and America's thirst for power. We are situated on the same planet and everybody is on the periphery of that planet. This should bring about immense relief to the soul of man. A person who is not sitting in the middle, cannot be encircled. The Germans have suffered for the last fifty years from a terrible fear of being encircled. He who is not sitting in the middle but on the periphery does not have to take himself so seriously. Occasionally he can sleep, occasionally he can quite forget his world mission, he can wake up to it again, for the planet is so kind as to be constantly turning, and to give each of us a chance to get the spotlight. For a time, everyone gets into it. Such is the wonderful feeling of relief which will be produced when the peoples learn that they are really living on a planet. Learn that is, not in the sense of school-book learning, just because Mr. Copernicus discovered it in 1543, but in the sense of Vilhjalmur Steffanson, as something we apply in daily life which means getting into the pleasant habit of discovering ourselves changing form and place upon the planet.

Thus, could a planetary order arise from the World Civil War, the result of the two World Wars would be the transformation of the earth from a mere world into the planet Earth. Then it would suddenly be possible even for the Germans to find a proper place on the earth, as the rotation of the planet itself would insure that there would be no more fixed positions, that everyone, as I said before, would get his chance. Then we would be finding our way out of the lost battles of the World Civil War. We should not be surprised if this transformation takes forty or fifty years. Imagine that as recently as 1893, the German Reich would have loved to have acquired a few islands in the Gulf of Mexico as a colony. That sounds ridiculous today, but it happened only seventy years ago. They already had steamships and telegraphy, but it was an odd idea to start a new German colony before the gates of New Orleans, or Florida, or Washington. The length of time needed to iron out and smooth away one single mistake in world history seems to me to be decidedly underestimated. The fact that I began with the quote from 1929: "*La guerre ce sont nos pères*," should also warn us today against believing too quickly that this period of cold war is over.

There may be a certain wisdom in the fact that the souls of men have been involved in war this long — from 1914 to 1965 it has now lasted — for that has forced more than one generation to look at these words "World War" carefully, and to uncover for themselves the secret of the transformation which has been laid on us by the word "planet" instead of "world." When I grew up you could still refer to the world and your native country as opposites.

You could sing praises of home and have contempt for the world. While visiting a backward looking family, I recently encountered just such praise of the homeland, just such contempt based on a poem by Theodor Fontane. So I had to say: "That's not how we men live nowadays, we aren't allowed to know beforehand what will be home, and what will be world. These two honorary names have to be constantly re-conferred. For where our fellow men fail, and do not live up to our trust in them, the word "native country" can no longer assert its rights over us. So in our era it is not so simple any more to separate the world from home geographically. My grandchildren's homeland is more important to me than my own native land."

The word "planet" should perhaps give us the hint that on the planet we are called upon constantly to discover new pieces of mere world and new true homelands. The name "planet" contains both. This dear earth is both homeland and at the same time very strange world. And somebody who thinks he or she can recognize homeland in the house which he or she is renting may be most mistaken. We spoke earlier about the fact that none seem to live further from one another than the residents of the same apartment building in a big city. Why? Well, one reason is, that we have to free many of our energies to be expanded on the distant world which also demands some solidarity. When I was young I was astonished when the miners in the Waldenburg emergency district or the farmers in the Silesian county of Neusorge knew more about world economy than my educated colleagues or the politicians, or the priests. The miners, for example, knew that their wages, the level of their wages, and the profitability of their work were dependent on what was happening in Australia, or on how much coal was mined in Yorkshire or Wales. The educated people did not know this. They were still trapped in the romantic notion that "home" was something geographic, depending on proximity.

The workers and farmers, however, - today they are even more aware of it - know that the price of eggs is determined by the world price of eggs, and that their work is in no way blessed because it is accomplished in their particular corner of the world. It has to prove itself on the planet. We educated people are left behind. For much too long, we historians, sociologists, lawyers, philosophers, literary historians, and professors of all kinds have been dragging around words which were eliminated by the two World Wars. We retreat in the face of the planet. The average man experiences the fact that he lives on a planet every day. Yet we are still talking quite modestly about the World Wars even about one single World War. We still distinguish between the so-called second World War — or a possible third — and the first World War. As far as I can see, the second time it was one and the same thing. In Hamburg my most German of all German friends cried when the Paris she loved fell in 1940. Anyone who was not sad then had already become brutal, a mere Nazi. Even a nationalist like Gerhard Ritter wrote correctly in 1964 that the Germany which he had loved died in 1917 with Bethmann-Holweg.

But even that is over: the unnecessary, superfluous second World War is over. The German educated Philistine and the German fleet of William II and the World Wars I and II are so far behind us that in 1939 already the leading European economist wrote to us in USA: "Greetings from the Continent of Europe which has a wonderful future behind it." An apparently later event can very well in the face of God be much earlier. Madmen live time backwards. That happened in Germany for example from 1933 to 1945. It was neurosis which caused Hitler to extort 1914's two-front war again from 1933 to 1945.

A very different assignment is before us today. Each fragment of planetary mankind has to watch out that it does not fall off this furiously spinning circular skeleton. Gone is the dream of some land in the middle which the Chinese held on to for so long. Gone is the fear of being encircled which helped the Germans develop their fixation of being in the middle. Throw all your world maps with their Mercator projections from 1558 into the fire. Then, at the time of emperor Charles V, Germany was indeed drawn in the exact middle, and that produced an idée fixé in the school books. You and I, we constantly suck in half-truths as children, in elementary school, high school, college, and during the wandering years. They are in no way untrue, just as Grimm's fairy tales are not untrue but they have to be completed into full truths by our experience. Otherwise the half-truths will let us become possessed. Tribal Germans, Christ, Moses, King and Country, "undefeated in battle" will plague us as half-truths as long as the eternal sophomore inside us is not kept within bounds. The Mercator projection of the world with Braunau (Hitler's birthplace), Linz and Passau in the middle of world history is just such a claim, believed from 1555 until 1914. The World Wars were required to give it up. Good geographers had long since recognized that the Mercator projection was outmoded, and had replaced it with better ones.

Just because we can put our own country in the middle at school, does not mean that this is its real position. The nationalists do not make this claim, but they do say: we will force this situation into reality. We will put ourselves into the middle. The result: they go mad. The Americans are hated for being too rich. The Jews are hated for being too regimented. And during this St. Vitus dance of mankind every hated group carries the mask of its sovereignty and its right to self-determination like a little crown on its head while claiming that they are standing as solidly and unshakably as German oaks. But "we should be shaken" — that is the beginning of being human. Anyone who cannot cry is not human. The folly of having to remain unshaken and of being in the middle are one and the same madness. Perhaps the Mercator projection world maps of 1558 with Germany in the middle laid the foundation for this.

When Bismarck's Reich ended in 1918 two people died for it, sealing the fact that the time of the Kaisers was over. One was a man who said "It wouldn't be nice at all if we were to come home victorious through the Brandenburg Gate." The other equipped Germany for its true future, exporting for the world as a whole. The first was Walter Rathenau, the second Albert Ballin. On November 9, 1918 Albert Ballin shot himself in grief, because his friend William II had made him wait in vain for four years without asking him once, how one might conclude peace with England. Ballin knew well that little Germany had no future without peace with the Anglo-Saxons. A country without a future sentences its best citizens to death. Albert Ballin died an honorable death. He did not despair over himself, he despaired over the future of his country.

The other man who died, Walther Rathenau, was never able

to make decisions. We know that he appears in Musil's "Man without Attributes" as a man with all attributes. He was able to do everything, and knew everything, and had everything, and was everything, and yet he was a poor devil. He has entered world history in peace because he accepted to be murdered. Walther Rathenau, it is astonishing, knew that one day he would be shot in his open car on Koenigsallee in the Grunewald in Berlin. Despite this, he refused to accept police protection. He knew what was in store for him. He felt as if released. He went to death like a soldier who goes to war in fulfillment of his existence. With this act of faith: "Yes, I will let myself be shot," he did a great service to the German people. He demonstrated the acceptance of the defeat of Germany. That was to be a good German between 1918 and 1945. For in 1918 the energies of the people were used up.

While Ballin demonstrated the opposite, despair of old imperial Germany having gone astray, Rathenau was brave in saying "It is true that we have lost the war, but exactly this was necessary, it made sense, and we must all continue living out of this insight precisely." At that time, 1918, 1919, 1920, 1921 there were very few who were prepared to take upon themselves publicly the loss of being in the middle, although all knew that it had happened. I don't particularly respect Rathenau for anything that he did in his life, but I respect him for the courage which made him say "Someone has to stand up and publicly state that the deadly future of that soulless man, Ludendorff, is really over." For that reason he became Foreign Minister, and that is also why he accepted the Treaty of Versailles, saying "the great power illusions of the European nations are over; we have all been conquered." That is how time is created anew - but not by the Ludendorffs and de Gaulles. If time is out of joint then everything moves toward destruction. Then there is no more present, since there is no partner with whom I can converse, as I have done hitherto with Mr. Smith and Mr. Enslin with whom I could separate the past from the future, and argue about it. When this is missing only the sacrifice of life can help. Ludendorff with his blunt soul could not make this sacrifice. So Rathenau outshines him. For despite all paganism it takes more courage to admit to defeat than to deny it.

Now I have to lead from Ballin and Rathenau to two men of the second World War. In Ballin we have the recognition of the achievements of the old Reich and the love which would rather disappear with the past than continue living. In Rathenau we have the faith that it is worth sacrificing your life to disclose their own future and the end of the past to your contemporaries. It looks as though both died in vain. They were both Jews, and the Germans said after 1933 that it would be better to live without Jews. With the result that once again death had to bring about the balance of threefold time. I will name two victims. One is Claus von Stauffenberg, the other is Helmuth James von Moltke. They too died differently, although they were both executed in order that it could finally dawn on the absolutely obedient lamb-like citizen that even the state itself can do unspeakable wrong. But they were executed on different planes, so to speak. Stauffenberg was a patriot, a nationalist, similar to Ballin. He shared the conceptions of absolutes, namely that the nation should embody everything noble. He fell with the cry "Long live Holy Germany." Not everyone (for example myself) may like this coupling of a worldly country and things sacred. That would come close to being blasphemy, were it not paid for with life. The risk of life may well let us be silent. When you risk your life you obtain rights. Like Ballin with whom I would like to compare him, he loved the good old, sacred German things, and tried to prevent their disfiguration, dishonor, and disparagement by the dregs of society, dying for them. But his goal remained the narrow German Reich.

The current division of Germany and the whole world situation does show us that he did impede this distortion as much as he was able, for one just man largely acquits Sodom and Gomorrah. But the inner future is not with Stauffenberg; that future is more with Helmuth James von Moltke. For he said "Don't murder Hitler, that is not important, he must lose his war. But after the war there must be men who can renounce this injustice, and we have to renounce it today in order that our renunciation will be believed tomorrow. Nowadays it is more important to be a Christian than a German under the swastika. For in the realms of this world, paganism has outlived its time."

In a legal memorandum I once said it like this: At the moment of their execution, Helmuth James von Moltke and his friends were the only legitimate government of Germany. For criminals cannot constitute a government, and at the moment when these men fell, crime ruled supreme, as far as it can rule. It may have the power to rule but that does not mean that it is legitimate. The epoch for the coming European generation — not only the German one — depends on our admitting the legitimacy of these victims of National Socialism. These victims of National Socialism provide the legitimate names for true humanity during this epoch. Just as the prophet Isaiah or the prophet Jeremiah are the only ones who count in the Jewish history of their own time, while we can in good conscience forget all the other Israelites of their time. Just as the prophet only represents his time in the whole rhythm of times, so these victims of National Socialism represented the justice which is part of every moment of the presence of God in the world. "When men fall silent the stones will cry out," for God must be present. If stones do not cry out then victims must speak, protesting against injustice. Much depends on recognizing that the time, the threefold time, can conjure forth its future even in the midst of downfall, even there where it seems to have no power. In 1944 no one recognized these sacrifices, and yet they laid the basis which has made it at all worthwhile to talk about a German nation today. If these righteous men had not existed in Sodom and Gomorrah, then Germany would have lost its history for good. For in every time destiny has to be created out of both past and future. The presence of God must be demonstrated, as otherwise the continuity of time is disrupted. And when this happens, some structure disappears into the abyss of the forgotten, and can never be brought up again. Only the ever present justice of God, together with his love and goodness let us share the continuing creation of history.

In the face of these victims, Ballin and Rathenau, and Stauffenberg and Moltke but also Leber, Bonhoeffer and many others, the decision rests with you to recognize the future even in defeat. These five or six people are more important than all the people who went along with events from day to day. Who asks after the mob's "patriotic" priests, of the Blombergs or the Keitels who scattered when Germany had to demobilize precipitously in 1945 and 1919, and people were languishing in hunger and misery? Only those who at that time stood up and determined the continuity of yesterday into the day after tomorrow have a right to give a name to their time. By throwing themselves as victims into the breach they gave their time its true destiny. They are the legitimate embodiment of their time. For out of mere occurences they made "events" for which they are personally responsible and for which they have suffered. Thus the time is set right, because their voices will determine its meaning.

I myself was a member of the legal faculty of Breslau which had to decide whether Techow, who had murdered Rathenau, should be allowed to enrole. It was a difficult decision. Rathenau's mother had forgiven the murderer in a famous letter. She wrote the letter to the mother of the murderer. My faculty was of the opinion that Techow could attend the university, but not study law. He had forfeited the right to become a lawyer. I consider this a Solomonic decision: Techow could be forgiven but was not to be left without some remaining stigma. Just as Golgotha, although remaining Calvary, had to be given up, and Peter went from Jerusalem to Rome, so some one who has broken the law as an assassin cannot afterwards become a judge, of all things, but he can become something else. He can be pardoned. He may not pass judgment himself, but may still do something sound. The effect of the event on the future, and the other way round, the effect of the future on the event is so serious that young Techow, in 1921 fundamentally a good boy, a misled, utterly undetermined person, won new insights while in prison. In 1933 he did not become a National Socialist either. He kept out of it because he had already — prematurely so to speak — been through the whole madness. After his deed he identified with his victim, Rathenau, sufficiently to realize that Rathenau had had something to tell him with his death. As he himself had caused this death there was a certain expiation contained in this. Techow's, the murderer's guilt was given meaning through the victim's willingness to suffer and by the greatness of his mother. The victim suceeded in redirecting the murderer toward the future, and as a result Techow did attend the university, even if he did not study law. The so-called worldly events are only the first half of the truth. Thereafter a soul has to perceive the meaning of the incompleteness and to complete it into a whole; so the mere half-truth of spatial shards can be healed into the full truth of the hour in real time.

World and soul join one another, that is they must join in order to provide meaning. For the world is meaningless. The world would just be uncreated chaos were it not for those who stood up with their lives to provide meaning; every time anew it would become an uncreated chaos in which speech decayed and every border became insurmountable. Chaos does not precede God's creation. No, chaos occurs when we little devils abolish God's word.

Overcoming of a border in not only a matter of guards, police, and passport inspection, but will probably consist of men talking freely and openly to one another. That is the decisive and always mortally dangerous step. When we think of abolishing borders we think of things that are too high-flown. It is something quite modest. We men build borders when we stop speaking to one another without reservation. We eliminate borders where we begin to speak as though we had no secrets from one another.

This formula sounds very simple but it contains enormous riches. We have to unfold these riches. In wartime we refuse the enemy greeting and answers, speech and response. If there is still talking back and forth as there was between the Homeric heroes, then the war is not quite war. When the soldiers got up to speak with their enemies over Christmas in 1914, their superiors received a deadly shock: the continuation of the war seemed to be threatened. It happened to me too even before the 15th of August 1914, right in the middle of the invasion of Belgium. A troop next to ours foolishly had broken all of the bottles in an apothecary's shop. When I arrived it was too late. In my memory, it seems as though a million shards were lying about there. But in the middle of this sea of shards lay a missal unharmed. I was so moved that I took my visiting card and on it scribbled a few words of sympathy to the owner of the missal, consoling him about the destruction of his property. Obviously I was driven to speak to the man concerned. I left the card inside the missal. This apothecary from Marchiennes wrote me in 1919, and I was the first German with whom he spoke after the first World War. I want to use this example to show that war stops speech; and we have cause to fear the interruption of speech more than bombs. Until 1918 the honorable tradition of transferring from speaking to silence to speaking again was functioning. Hitler the exterminator, ended that. He liquidated that peace for ever.

4. Nipped in the Bud

Now I should speak of the beginnings which have been nipped in the bud during the long waiting period between the first World War and now. We have made two points: first, that "war" and "peace", deadly earnest and parlour games, have become jumbled because men are saying that war cannot be conducted anymore. The Bolsheviks and the Pope agree on this. On the other hand true life must always emerge from deadly even bloody earnest. No one knows as yet how we can demonstrate such seriousness now.

If all borders of human achievement must stand unchanged, "where can there still be births"? we asked. Up to now the deadly earnest of war has always been the carrier of life.

The second point was that "world" and "home" cannot be distinguished from one another as easily as before, because both your home and the world have come to lie together on the rotating planet Earth. The globe stood before us hitherto and we children could spin it with our own hands. But now it has grabbed us, and we are beginning to notice that we live on a planet. No one is in the middle; nothing is mere world, nothing is mere home. The ham radio operator in Chicago may have to free his own neighbor in Tokyo.

And we added a third realization to these two points, which I would like to expand a little, namely that a border can run through one's own people and had begun to do so in the case of the workers' movement a hundred years ago — because labor was not spoken to. Its face remained unsightly, so a border appeared right in the middle of daily activity. This border seems insurmountable and indestructable to many even today. An engineer does not talk to a worker; instead of looking him in the eyes, he looks only at an efficiency study of his movements. This third point should help us explain why mankind has become restive since the first Russian revolution in 1905. At home and abroad people have tried to replace the seriousness of war (which was becoming impossible) with a new seriousness of service.

I must tell you about these attempts, these buds, although in retrospect they often leave a sour or even bitter after-taste. For example, I recently read about a rally of a German youth group which refused even to consider a work service organization for German youth, although they had to admit that just such service would be necessary in the developing countries. The German "Arbeitsdienst" under Hitler has left an anxiety behind it, and people exclaim nowadays — probably without experience in the pre-Hitler work services or even in Hitler's work service itself — "Anything but that again!" Thus very many anticipatory and important efforts have been sentenced to silence, to being forgotten, and it is questionable if anything original can be produced to replace them.

These older responses to mankind's new emergency of being shut within its boundaries were probably profound and certainly very well thought out. And perhaps they were correct even though they are being forsaken by contemporaries. The failure of first attempts does not disprove them. Anyone who ridicules them is a slave of success. I will simply say that I have been in a position to recognize the difficulty of dialogue between labor and other fellow citizens in society, both at home and abroad since 1912. I recognized that this dialogue had been so completely destroyed that it could not be revitalized by discussions and entertainments and play acting or choir groups.

Why is the post-Hitler Youth Ring against work service, and why will work service corps necessarily come into being anyway? Well, in the red-plush society, the Good Society of 1875, the playful sort of small talk and entertainment called discussion had replaced the serious Word which truly opens men up to one another and "unbounds" them. I still meet people today who do not know that only they who have worked together can pray together. The ancient saying "ora et labora" (pray and work) is a fact of human life. You cannot just "start talking" to use the current jargon. People have something to say to one another only when they belong to one another or when they have toiled together. The sort of speech I am going to be talking about is not produced in club meetings or cheerful get-togethers or small amateur plays. After working we can speak together. Because we have suffered and haggled together over the work.

The thrust of the work services which have been recommended, started and tried since 1910 - and which are still suffering today from their misuse under the Nazis - these work services were not primarily interested in the work as such, but preparing human conversation. Only people who worked together really have to sing together afterwards. Work is probably only the minimum requirement for this. The people who have cried together or worried together can also speak together but the simplest and most universal prerequisite is human labor. Work as a prerequisite to the Word has been the secret of all the attempts from 1910 on to cross borders entirely seriously without bloodshed.

This kettle was boiling everywhere. As of 1905 "world" politics appeared unexpectedly in the thoughts and expressions of the powerful, and the expectation of many national revolutions was replaced by the expectation of a world revolution. This world revolution then took the form of two world wars with an armistice of twenty years between them. It became clear that the world economy would lasso the people one way or another just as it is obviously doing today.

This cold threatening world challenged us. We had to find out whether we could confront the mute world with a harmony of the spheres, a planetary order. Because no man and no people can endure a mere world. Under the worldwide and yearly threat of impending war I too became aware of a future beyond the wars in an unexpected way.

The idea came to me in Heidelberg. I was writing my professorial dissertation there in the winter of 1911-12. An acquaintance, Werner Picht, invited me to come to the so-called Max Weber seminar, a group of Ph.D.'s and doctorial candidates. He was going to speak on the English "Settlement Movement" (Toynbee Hall), the community house which brought the rich and the poor together to meet as human beings over tea. We listened to his talk. Now Eugen Levine was there, a tried and tested Marxist who had gone through the school of czarist prisons, and who later gave his life for the revolution. His reaction was "As long as the bourgeois children do not play with the workers' children I won't believe in your human reconciliation and your brotherly attitude." And thus cornered, we Germans were put in a position to say "Let's create a work service corps." I wrote the Prussian Ministry of War, saying that there were so many one- and two-year recruits freed from military service for ridiculous reasons of health. They should all be collected together in a social service corps where they could practice living and learn how to live.

I was able to explain that memorandum from 1912 once again in 1956 to 300 officers of the future Bundeswehr (West German) Army in Sonthofen. General Speidel well understood that such a corps would take a moral burden off the (West German) military. A social service corps would have offered moral support to the remilitarization of West Germany, a very premature act.

And I am still of the opinion that every service corps, whether national or international should measure itself by the recognition on the part of the military, that such a service and service in time of war are two sides of the same coin, and that war now more than ever needs to be ennobled by training warriors to behave humanely. Most people consider such a stance crazy. Most such budding efforts have not been planned in harmony with the military but in opposition to it, by pacifists. A whole row of such attempts have arisen in Switzerland. After the first World War Pierre Ceresol, the good son of a Swiss officer, tried to develop such an atonement service corps for the whole world to iron out the horrors of the first World War so to speak. In Germany Alfons Paquet, Martin Buber and Florenz Christian Rang called for a reconstruction of France by peaceable means in 1921. As far as I know, nothing ever came of this. After the Second World War there was a whole list of such organizations, of which the foremost in Germany is the Reconciliation Service Corps (Aktion Sühnezeichen) which has, as far as I know, successfully done service in parts of the world like Greece and Israel, and which is now trying to do service in Russia,

Yugoslavia and Poland as well. The Swiss "Zivildienst" which has been doing things of this kind for decades, is active in North Africa. In America the Quakers have been organizing camps in which young men and women work in Mexico, or in the Japanese sections of California, or in disaster areas. They repair school houses and build streets, for instance.

These Quaker enterprises are irreproachable in their spirit, but they have one defect, namely that they are vacation enterprises, and things that students do during their vacations just are not entirely serious. Vacations are treated as pastime, and so the Quaker camps run the risk of having the transitory character of "one try doesn't commit me" as prevalent as deadly-earnest is in the army. Even a soldier's peacetime training in preparation for war is bitterly earnest. Fifty years ago a francophile pointed out to the English that they never had to undergo the bitter duty, the shocking seriousness, of losing a real chunk of their lifetime, and of being innoculated with a completely strange order. Every Frenchman, however, experienced this in his military tour of duty. As Belloc said, "Three years are no child's play". I myself have spent six years in the Army in peacetime and in wartime, and if I count everything together, another three years as a volunteer work-service soldier, when I actually worked with a shovel or similar tool. In the First World War, when the morale of the troops began to sink, I was allowed to organize a work-camp for the boosting of morale in my division right behind the front at the divisional headquarters in Champagne. That was in 1916 — I have also publicly described this — and it was my first experience with work service camps.

It is quite important to recognize that there is a border in the Army also, between officer and man, similar to the border between engineer and worker in the factory. And it is just as necessary to celebrate Saturnalia there — that is circumstances in which man and man, not officer and man, live together — because political parties and social groups are much too directed towards certain interest groups like workers and management. The layman and the priest, the officer and the man, as well as the engineer and the businessman and the workers, should learn to live together incognito, anonymously. Working together lets everyone forget the ranks, the separations, and the hierarchy, and lets dignity, authority, and competence be revealed anew.

I tried to reproduce the experiment of 1916 in various forms after the First World War, and it is important to give you a small taste of the misunderstanding that is prevalent. Count Keyserlingk (a quite well known German philosopher) had to say about my proposal of a service of several years for the students of the technical college in Darmstadt: "Anything that the students can't learn from me in half an hour over a glass of champagne, they won't learn from you in a year either." This brings us to another serious aspect of the willingness to serve. You can play just as well in half an hour as you can in a year. But you cannot develop real seriousness in half an hour over a glass of champagne. As war must be replaced by something as earnest or nearly as earnest, I must insist that it is necessary to sacrifice some period of time, a considerable period, a chapter of your life. The brilliant Count Keyserlingk could presume to believe that a flash of genius can be just as effective in half an hour as in a year of contemplative suffering or silence. He misunderstood the contest between seriousness and play. He was a thoroughly playfull man all his life; he was never quite serious about anything.

My attempts continued, and we had the greatest success in Germany with a group of friends in the "work camps for workers, farmers, and students," which started in 1926 in Silesia. One hundred young people came together and lived there for several weeks — one group stayed several months - in order to consider together the serious grievances which were prevalent in the depressed areas of Silesia then. The example caught on and ran like a forest fire throughout the country. In 1929 such mixed work camps existed in perhaps 15 German provinces. Our Silesian example had proved irresistable. To be sure, many of these camps shared with the Quaker camps the defect of lasting only about four to six weeks, and of avoiding the harsh seasons of the year.

But they had an advantage which I cannot recommend highly enough to the reader at a time of economic prosperity. These camps were held in a similar boom period, in times when no worker was unemployed (or had to be), in times when we had to drag each participant painstakingly away from his profitable occupation to come to us in spite of prosperity. Here we can learn about one particularly great danger along with the other ones surrounding such service groups, that people do not start doing necessary things until the country is up to the ears in unemployment, when it is no great shakes to say something must be done for the unemployed.

That is when the monstrous mistake is made, of massing the unemployed people together, of leaving them alone with themselves. Everyone who is unemployed is indeed stuck into a work camp or sent into a service corps. But the other men, those who have their lives before them and are certain of their future are left outside, and the misery of the unemployed is infinitely increased because they are isolated.

When I arrived in America, they had just finished copying this mistake. President Roosevelt had founded a Civilian Conservation Corps. I would not deny that at the beginning this corps was very useful and necessary, in fact indispensable, in bringing the unemployed youths from the big cities to order. But the basic flaw of the German Volunteer Work Service Corps of 1931 and its imitation in the USA was the massing and isolating of the unemployed among themselves. The last thing that an unemployed person can take is separation from society. For the very reason that he is without work, he needs more contact with people unlike himself than the employed does.

It seems dangerous to me that no one nowadays seeks to separate the two situations: economic prosperity with its lack of labor — when we had to drag the people from the well-paying jobs to let them experience serving instead of working — and the depression, when the despairing masses had to be taken off the street - so that people would discover that they belong to an ordered society in spite of unemployment. The factory and the office provide us with order. The reincorporation of the unemployed requires the discovery of a further relationship. The years between 1929 and 1933 showed us what a gap opens between the workers and the unemployed. And it is much deeper than that between entrepreneur and shop steward. As the experiences made with the unemployed have disappeared in the boom period, I would like to point out a difference which is of importance all over the globe; an unemployed person becomes "nervous", uncertain, hysterical. He gets on the skilled worker's nerves. An employed person looks away from the unemployed. *The worker is scared of a person out of work.* He fears him because he represents excessive supply and may undercut wages.

As we like to imagine nowadays that an economic crisis will never recur, the reader will perhaps be yawning at this point. But in the developing countries contact with the industrialized world is producing new unemployment, and so the Germans' experience is of interest. I myself have been the victim of the friendliest form of forgetfulness which is being dumped over the scenes of industrial history these days. May I talk about it for a minute?

My friend Hans Thieme wrote about memories of Silesia, and wanted to honor me in them. So he mentioned that I had tried to help the unemployed in 1930 with work service camps and that I had thus become the forefather of the Kreisau circle (resistance group against Nazism). But the true story falls apart into two, in retrospect overlapping, but at the same time entirely separate periods of time.

From 1923 to 1929 I took up arms against the tendency to forget the World War. I believed that it was not over yet because the German upper classes were lying to themselves. I went begging among the returning workers, farmers and students who were full of the pleasures of life, for the seriousness of a common work service for peace. My first prospectus was called "People in Industry." It developed the theme of my 1912 paper and my wartime work in Champagne and disappeared in the first of the prosperous years, 1924. In 1926 I was able to print the documents from the years 1912 to 1926. With the dust which they stirred up I was able to win over students (drunk with the youth movement), young farmers (drunk with nationalism), and young workers (drunk with socialism) for the first Silesian work camp. At its culmination all the prominent men of the province came together for three days, Catholics and Protestants, industrialists and agriculturalists, Church and State. Thanks to Heinrich Brüning, our Reichstag representative for middle Silesia, the twenty-year-old Helmuth James von

Moltke (who was hanged by Hitler in 1945) was able to get 6,000 marks for the camps from Hindenburg's (the President's) pocketbook. The economy was so prosperous that we had to convince each member of the camp painstakingly to take part. By 1929 this political pattern had found successors in fourteen different parts of Germany. People living during the current boom period would like to get by without living out such a liturgy.

In 1930 the depression and unemployment hit. The lines outside the employment offices became the spiritual cesspool from which the brown shirts necessarily emerged. I never had a chance to communicate the enthusiasm of the first camp for farmers, workers, and students to these despairing masses, friend Thieme. Rather I went to the governor of Silesia and warned him, "Can't you see this avalanche of pollution rolling towards you? Yet these masses are being bred by your method of providing unemployment compensation." And then I discovered the helplessness of this old union man: he was horriffied by the unemployed. He felt incapable of getting closer to them. The villas of the rich in the southern part of Breslau were standing empty by the dozen. I proposed to take fifty villas, to man them each with 8 - 10 of my students, to divide the unemployed people into fifty groups and to pay their unemployment money through the students. I promised that these fifty groups with their ten students would coalesce and become friends. In this way a human order would replace the degradation of the unemployed lines. The governor just laughed at me. And that was the limit, my dear Thieme, of my efforts on behalf of the unemployed (on the other hand, my last work camp like those during the boom of 1924 to 1929 was held in October of 1932 in upper Silesia).

The gap between boom and depression in Germany divided two different times. Both happened neatly separated. What is different about India, or Burundi, or the Philippines? Both boom and bust are happening at the same time. Although this modest book does not have any cure to offer (if only because each country from Costa Rica to Somaliland requires something different), the following always applies: industrialization must take two things into account: (1) Unemployed people must "belong" and be able to make friends in spite of the fact that they are out of work. (2) Employed people have to stick their noses beyond the borders of their jobs, physicians as well as lathe operators, ministers as well as alpine farmers. Each fellow man must have his off-work - and Sunday face brought forth. Only a person who alternates between the two faces, the working face and the celebrating face, can be called our fellow man.

It may not always be possible to combine measure one and measure two. We tried to do so in the USA from 1939 to 1941. Brüning's Work Service Corps of 1931 (which the Nazis later spoiled), was reconstituted for the eleven million unemployed people in the USA in 1933. It was called the Civilian Conservation Corps. The Corps began full of vigor, with the fabulous American ability to improvise. It was murdered a few years later, however by Saint Bureaucratius.

In 1938 I convinced a few outstanding students to enlist in the Corps — which was after all only five years old — in various parts of the country. This worked out, and it caused the revelation of rather hair-raising conditions in some of the camps. The bored boys — between 17 and 22 — ambushed cars and on payday it was necessary to have a revolver on the desk to insure that no one stole the payroll. One educational officer — often reserve officers from the Army - told me personally (and to his credit one should add that he was ashamed) that he paid cash to the fifty inmates of the neighboring camp so that they would condescend to play games on Saturday and Sunday, which they had free.

And the saddest of all: at communal meals the rascals grabbed the nicest pieces of food away from their fellows. It is an important truth that the origin of every community is rooted in and should be rooted in the meal. Animals gobble their food. Individual men eat. But a meal unifies to such an extent that each man offers the best bits from the communal store to the others. They have to, as otherwise the peace will be broken.

One can see here that peace does not just mean that we do not beat each other up, but rather that peace changes the participants into partners. Otherwise it would just be an armistice. The childish delusion of the educated that peace is something natural, disguises the contrast between armistice and peace. Without being joined by the employed people, the unemployed could not and cannot make peace with society and could not even offer peace to one another. One hundred unemployed men in an isolated camp - I found this also when I studied the work service in Bulgaria in 1927 — are, because they have to remain among themselves, more of a Revenge Corps than a Peace Corps.

This glance at unsuccessful work service groups adds an important element to our understanding. One hundred fragments will not make peace with one another just because they were all broken at the same place. Shards can only be made into a whole vessel when they are glued to complementary, matching pieces. We gathered about fifty college students together with fifty unemployed men, because massing the unemployed together had proven only to heighten the breach of peace. This transformed the emphasis so thoroughly that no one needed to revenge himself. From Camp William James, so we hoped, were to come the new Corps leaders whose training President Roosevelt entrusted to me. Thus we were able to weed out boldly a mistake which is still present in the heads of most unconcerned people, and which again might become deadly delusion as it did thirty-five years ago.

We invited two of the sons of William James to the christening of the camp. Through them, this man, who in 1910, prophesied a heroic Service which would replace the warrior, gave our camp his name: Camp William James. The Japanese attack on Pearl Harbor on December 7, 1941 put an end to this camp for training peace service corps leaders.

No practical churchman or statesman has yet thought through the full dimensions of the accomplishments possible for such a peace corps. In a pamphlet published in 1932 ("Work Service-Military Service"), I tried to extrapolate theoretically the point at which people must arrive with advancing automation, and I came to the conclusion that only a peace corps can prove equal to future society.

In 1908, however, a radical spirit thought out a much broader future revolutionary change. The uncle of the London economist Karl Popper was an engineer, and published a paper in 1908 in

Dresden under the name Lynkeus. This paper demanded that all young men serve ten years, producing all the goods necessary for the life of the people. Thereafter they would be free to do what they wanted to do. He had figured out that this 10-year service would suffice to produce all of mankind's necessary goods, above all the daily bread. Here our undertaking has been turned upside down, so to speak. In place of a spiritual unbounding through communal work and the resultant conversation, Lynkeus-Popper wanted to call upon men to undertake a service which would replace the whole social order of private property with a public service in all countries of the world. It was a very intelligent paper, and I mention it here only to show that it certainly is one of the shoots today being nipped in the bud by the contrast between capital and labor, between the so-called capitalistic and the communistic countries, an artificially puffed up contrast. The bankruptcy of communism's sheer war economy is obvious; but perhaps Popper's thought might have a future sometime. The idea is not bad that if we could get rid of the military, the years of service previously spent with the military might be used to produce the immediate needs. I personally have no faith in Lynkeus' proposal. But it is exciting to see how, during the transition to a new world order as from 1905 to 1965, people in different corners and at various points have had to mull and are still mulling things over.

There is the fact that the failure of plans of such quality causes damage; when they are given up again, people become discouraged and say, "it's no use." I have been telling about all of the beginnings nipped in the bud; I have been telling about my own disappointments. They who see only the misuse of things which are right and necessary are often the very ones to obstruct them. Extraordinary faith and special patience are needed to repeat new forms, and so to force them into being, in spite of the unavoidable early mistakes. Our seriousness is first really put to the test by failure. Amazingly little is said about such endangering or damaging of future life by short-lived expectations and impatience. I have never heard a minister give a sermon on why failure is part of success, and why the only path to success must consist of telling the young people that a cause may become legitimate only when it has once failed. The work services cannot possibly catch on in a world in which Americans want to own part of Panama in perpetuity, and in which a Napoleon IV (de Gaulle) still dreams of the hegemony of France in Europe, where men in each individual state aspire to continue the vocation in which they were trained by their grandfathers.

The contrast today is very funny. On the one hand the young say: "The war is our fathers' business;" they are psychoanalyzed; they are born with silver airline tickets in their mouths and saunter through the world. On the other hand they leave the most pressing of the new tasks lying because their 150 year-old brain tracts may not be touched.

There is no reason to exclaim, "never again a work service," as the criers in the German Youth Club did, just because the first attempts were nipped in the bud. Perhaps we shall have to rename what we are going to have to do. But just because something has been botched does not mean it is disproven. On the contrary, the old, the decrepit, the reactionaries naturally try to grab cheaply a piece of the new, and send the water over their dam. We must clean the misused, we must renew it. But it is not disproven because it has been misused; coruptio optima pessima — the best is always first misused and this is the worst corruption. That means, however, the most difficult to understand, the newest, is most easily yanked back into the old, to help out the old, to fluff it up, to save it for a little while from total decay and moldering by apparently reforming it. This is what has happened to the Services which are to replace the military service. The national powers stuck them first into the outmoded political strait-jacket of nation-state vanity, and are still doing so.

Misuse never disproves the proper use. At the same time when the Caesars' let themselves be abominably worshipped as gods, it was nonetheless true that God has become Man. Today the sovereign states let themselves be catered to, although they cannot remain either sovereign or states. The truth however is that the new sovereign must be served instead. But that is not Mr. de Gaulle.

5. It Can't Go Slowly Enough

Our task will be to rediscover the right attitude towards serving seriously. All the talk about the developing countries is getting in the way of this, as is the talk about the aid which the more fortunate (that is, rich) countries are giving them. I don't have anything against aiding the developing countries. But if the serious service corps which I am concerned about is tacked onto aid programs, all efforts will be nipped in the bud once again. Ever since 1905, these seeds have been planted in vain and cast into a power-hungry, bellicose and exploiting mankind.

That is why I think that the talk about aid programs, the talk about pluralistic society, and the talk about the economy in general is dangerous. The economy does carry us easily into larger communities today, and we are right in welcoming the economic sector because it liberates us from wild nationalism by de-nationalizing us. But we will create too much of a dollar-democracy if we mean by "economy" just expansion, or the exchange of goods and services across national borders. It won't work for the simple reason that riches create enemies. Veroni, the Italian ambassador in Bonn, said: "If we continue this, our loans to developing countries will just make us enemies."

It won't be possible to tell the difference between self-interest and service if we mix up aid programs and willingness to serve in Africa, be it with the Peace Corps, with the Reconciliation Corps (Aktion Sühnezeichen), with Crossroads Africa, or with the Swiss and English Civilian Service Corps. (I want to praise the English one in particular because it is so wonderfully small, with 96 members at present, each one of whom does first-rate service.) Aid programs do employ people, and build them lovely houses. Two million marks were spent in a certain West German city on visitors from developing countries — that is, they built enough bathrooms. For the visiting Undersecretaries, a private bathroom for each was built, just to make sure that they felt completely at home in Germany. These good intentions did not quite work out. One evening, one of these Undersecretaries of the Bathroom went back into the main villa and asked for a cup of coffee. "I'm awfully sorry," said the housekeeper, "but we have only one coffeemaker to make coffee for all 90 visitors." The Siamese gentleman would have preferred a personal cup of coffee to a personal bathroom.

I made a counterproposal, that they should arrange to give as many foreigners as possible the experience of living in a German village with a good village schoolteacher. People from abroad would learn far more through the spirit of peace, patience, and freedom in a teacher's family than they could ever learn from all the bathrooms and reception rooms of the Villa Such-and-Such. Riches create enemies. Doing good deeds may give the good deed-doer a nice feeling, but it is an old rule that the recipient of mere charity often tries to revenge himself for it. Charity is dangerous, when the rich come to the poor.

I can only recognize as service the action of a person who knows himself poor, and who serves in order to raise himself and become more than he was. Now how can a man from a far-too-rich society which offers all possible consumer goods become so obviously poor that he will be loved for the services that he undertakes, instead of being hated for seeming to show off his excessive weight and power? That is a hard question. It is of no help to the rich person that he means well. The misinterpretation is not appearing in his soul, but in the recipient's who cannot separate the wealth of the giver from the giver's poor, fearing, searching, and quaking soul. This has always been an important question and it is nothing new in world history.

Becoming poor is perhaps even harder now than it was in times when you could obviously tell rich and poor apart. Nowadays a sort of confusion exists about who really is rich, who really is poor. At the moment, no one is poor, since we happen to be in a boom period. This is using the expression "poor" in a poor way however. Who is poor and who rich will become apparent only in the course of a long life. Therefore, at the moment, forming serious service organizations which would draw as much out of men as military service once did, poses a difficult question. It has been called the burden of plenty; the affluent society is leaving visible traces on our souls and on our bodies. We are not helping the underdeveloped countries when we give them part of our wealth. We are just busy creating customers. The plant which is being built in Pakistan will soon rust because the necessary oiling, care and cleaning are stopped after only eight days.

Or consider the chief's old hut standing next to the palace which he has had built for himself by Western architects and engineers. After four weeks he has already decided to move back into his hut and leave the palace next door as an empty status symbol. This story from Africa is true and can be multiplied many times. The goods of the world are being poured into these new political units so quickly that they don't reach the right people. But goods only benefit the right people.

And so we should pay some attention to the wavelength to which the souls of real men respond. When we turn on a television set, our eyes can keep up with things, but the person to whom the eyes belong cannot. I would like to examine some significant experiences in this light which, although determining Germany's fate, have apparently remained completely unrecognized in Germany itself. Black chieftains are not the only ones who learn nothing when they are made to learn too fast. All peoples are like that. In 1919 Foch said, "Twenty years of armistice." In 1945 Roosevelt said, "This time we will have to wait for at least five years to conclude the peace." The Korean war of 1950 has proven him right. While in the past, hundred-year and thirty-year wars dragged on, modern technical war is concluded so quickly that the peoples cannot possibly keep up with it. It was said of the battle against the Huns in 451 that the dead had continued fighting through the night. Now the Germans technically had to stop fighting in 1918, but the war was not over with that.

This will help clear up a gruesome misunderstanding from the year 1945. At that time many Germans naively expected the

Anglo-Saxons to start fighting the Russians with them. This shows a false and mechanical concept of war. The English and Americans were not dead machines. The Americans took from 1941 until the end of 1944 to get their hearts into war. It was not until the Battle of the Bulge at the end of 1944 that the American hearts became possessed of the passion for war which the official German psychologists thought they would have at its outbreak. At the end of 1944 the passion became so great that even my best American friends suddenly shunned me as a wicked German. Until Christmas of 1944, they had never given thought to the fact that I was a German. Thus in 1946 and 1947 the Americans were in a much more anti-German mood than in 1942, which shows that they are human beings, not tin soldiers. They needed time; and their hearts beat in an organic time with their own rhythm. Here is a lesson which will destroy politicians' concepts. The lesson is: the technical part of life moves ahead much faster than we can experience it. A war may be already over technically just when people are being passionately carried away by it. Hundred-year wars will not happen again, but there may well be 100-year cold wars between armistice and the conclusion of peace.

The times of peoples and the times of technology are different kinds of times. I have formulated the laws of technology like this: Every technological advance shortens time, widens space, and destroys a familiar living group. As soon as we understand that these three rules apply to the World Wars as well, we cannot be surprised that the technicians are having trouble coping with the developing countries. The three gifts of technology: the acceleration of time, the widening of space, and the destruction of the existing peace, must be offset, and they must be offset with the help of people who have nothing to do with technology.

Aren't the efforts on behalf of the developing countries self-contradicting? People talk about "developing countries," and in the same breath about plans to make them happy. Alas, a person who can be made happy surely cannot develop. On the other hand if you are literally "developing," then all we have to do is let development take its course. But we should not wear this pallid 19th-century mask of "evolution," and "development." "Development" is meaningless. Peoples can be ruined, empires may topple. Nations may go crazy. Developing is such a tangled affair that it would be better to keep our fingers out of it. We should not denigrate the new states as "developing countries"; development usually means decay. We might modestly take another step, however, which would assure them of our solidarity. As "developing countries" they are objects. Who dares to approach another people except as their "neighbor" (Biblical sense: see p. 9). A big steel foundry, complete with bordellos and bathrooms cannot be the next step then.

"It can't go slowly enough," said a farmer when I asked him how his village could be modernized. It can't go slowly enough? Since our technology has waged the World Wars faster than our souls can, which after all should and must experience in generations, the response, "it can't go slowly enough," I feel, stops sounding funny.

Anthony Eden, the former British foreign minister, suffered a heart attack in Cologne after he had dropped in on three European capitals one after the other. That was thirty years ago, and I inwardly doffed my hat to him at that time. Eden was no wind-bag like Ribbentrop, and no bureaucrat, either. Although he was certainly not a very gifted foreign minister, he did have a human heart, and it lost its balance in the collision of three political climates: Moscow, Rome and Berlin. The 18th-century German poet Lessing said that anybody who won't lose his mind in certain situations does not have one to lose. Nowadays this should read, a person whose heart won't go to pieces after certain flights, does not have a real heart. I wouldn't trust a man or woman like that. But the political public, that beast, expects us all to fly. They believe that faster is better. World politics are governed by this nonsense these days. What could be done? As people we are multiform; we can divide our labors just as stomach, heart, lungs and liver do. The foreign minister may have to fly. That would be bearable if other members of society insisted on slowing down as a counterweight to his lightning-like flights.

Really, if things cannot go fast enough for you, then the other side of the coin is obviously a world view according to which things cannot go slowly enough. Our planet Earth also moves both quickly and slowly at the same time. In the eyes of God, one thousand years are like one day, but to a child of man, a single day may stretch out unbearably in sheer unending toil. You and I, however, who carry both God and man within us, must embody both measures of time concurrently: we have to be partly flashes of lightning and partly snails. Every event contains the extremes of lightning flashes and snails. And for that reason we are burdened by both as if they were one. Cultivating the fast or slow rhythm of reality only contributes to bringing about the end of the world. So someone who wants to realize the dream of a thousand-year Reich before he is fifty, will necessarily bury his people in ruins within twelve years. Peoples are not flashes of lightning; they are snails, suffering from syphilis for three generations. They are not able to hand out charity, mercy and justice and show their humanity within a single day. They will reveal whether they have recognized the demands of the day and patiently lived up to them only in the course of a century. Those who do not fulfill the demands of the day, be they quick ones or slow ones, drop out of the Christian era back into prehistoric times. That is why Hitler was a pre-Christian; he could not wait for the blessing. The Fellahim of Egypt have been non-historical from Ramses VI to King Farouk, since foreigners were ruling a mass which could not make any decisions.

The oddest thing about having both the lightning and snail characters simultaneously is perhaps not so much their connection but their arrangement. If a child is starving and the Vatican starts to burn, the child is more important. If you would rather save the burning house like a flash of lightning, while the baby dies of weakness in the meantime, then you have forfeited both the baby and the Vatican forever. No one will believe you anymore. For the child is more important than the Vatican.

Considering that, the reader may perhaps be patient as I warn him that the developing countries are like the baby. The house is burning; revolution, war and starvation are flashing up. How can you even think of mentioning a snail's pace? Are you ridiculing us and all the experts from the World Bank, the international conferences and the United Nations? Speed is of the essence. The danger lies in delay. Otherwise we may expect a world-wide economic crisis and revolutionary outbreaks.

Impatience and indignation are understandable. And before I can convince men and women who are concerned with these new countries of the need for my snail's pace, I am going to have to ask the Westerners for as much patience as they will need to observe their own immense life span, stretching out behind today's installers, engineers and technicians: their own pasts. The word "Westerner" was an insult under the Nazis, but those same Nazis perished because they shared the indecent haste of the Westerners. Thanks to their ruin we have an incredible amount of time, even if the West believes it has no time at all. And so the West (that is, booted and spurred industrial mankind) is busy changing every day. It embodies progress, speed, movement; and that is why Mr. Mao who wants to outbid us, also speaks of world revolution. The poor man has to outbid us in talk at least, since the Chinese must catch up on two thousand years. Considering the Chinese, the honorable reader will perhaps listen to me, for I will admit openly, we will not remain ahead of them with Planck, Henry Ford, Rutherford and Einstein alone. The Chinese can buy these people, and that is why they have already produced the bomb. Those are ready-made products of the moment.

Christ is Lord of the moment, too, but he is Lord of eternity at the same time, that is, both man and God.

All of the speeches of the chambers of commerce about the developing countries read as though we, like Mohammed, believed in Kismet instead of God. In those speeches, as in Islam, God does just what He wants to; and we do just what we want to, failing regularly. Luckily Allah is merciful. So we are just barely there, just as the countries which Islam overran between 700 and 1900 are just barely there.

To us God has become alive in three Persons. That makes Him far more terrible than if he were only merciful, but also as powerfully creative as at the dawn of time. And the first thing about the living God's creation that comforts us, is its inexhaustibility. So if God has not only become world but become man as well, then His inexhaustible power to create new men can be soberly demonstrated by the appearance of men who have never existed before. We live in an economy thrice renewed in the name of the threefold God, and we are starting to meet the other half of our race to which this triple renewal has not occurred. Before selling them television sets, we should look at ourselves.

Europeans have taken the risk of a new economy three times. This happened the first time when the wandering Teutonic tribes were ready to settle down under the guidance of the bishop's staff. At that time the active part of the population, those whom we call workers today, were called *peasants*. They had the same significance that the workers have today. And those whom you could compare with the engineers were the monks under whose guidance the land was cultivated. The peasants and their teachers, the monks, did something very practical which we should give to the world again as a European export. They made the desert, the thickets, the primeval forests, and the swamps arable. Having made them arable, they did not leave the land to turn into prairie again after the first exploitation, as we see it done in Asia Minor or Mexico today.

It becomes apparent, looking at the migration from the East, at the reversion of whole countrysides to barrenness — as in Spain, Morocco, or the whole African continent for instance — that there is nothing self-evident about continuity in the cultivation of land, in not leaving it to lie fallow, in not moving on after exploiting it for thirty years. There is thus something about this European peasant's attitude which should be retained and passed on.

The crafts were the second economic step in European history. Under the leadership of the architects of cities and cathedrals, the abilities of the craftsmen were allowed to develop in the Middle Ages, regardless of how much time it took. Nowadays the crafts have fallen into disrepute, but German ability to compete abroad still rests largely on the skill and industry of craftsmanship. With so-called automation we forget too easily that reliable workmanship and continuous quality in a flawless product are in demand.

The new division of labor between engineer and worker nowadays, often obscures the craftsman's skills with which we are still familiar. The third epoch of European economy is marked by technology, as expressed in the relationship between worker and engineer as doer and leader. Peasant and craftsman have moved somewhat into the background.

The tasks facing the Europeans in the developing countries give full meaning to the heritage of these three steps of economic life. The developing countries should not be given technical progress and teamwork only, but as well, men who have retained in themselves and in their method of working something from all three steps, within ourselves. We must have some part of the peasant who takes it upon himself to settle permanently in an unproductive area. This time-honored virtue will be needed, in order that there not be only rapid booms and busts and rapid exploitation in the developing countries, but the construction of lasting institutions. People will pump oil out of the Sahara, but they won't open the region up the way it should be opened up: as a place to live forever. Whether we take Brasilia, the Sahara desert, or the Congo, whole cadres of workers are needed who will remain active there, generation after generation, whether they are physically related or not. Once opened up, a region must remain part of human society. Men must be given an incentive to stick it out in any climate, incentives like schools, church- and community-life. One cannot always count on migration and relocation.

The other special dowry which the Europeans still have to bring to the other peoples is reliability at work. Every worker may be worth his wage, but that is not the point. It is important that every piece of work receive all the attention it needs.

An engineer has to contain the heritage of the peasant and the craftsman. This is important for men who are being sent abroad to build a factory, for instance. An installer must carry the peasant's and craftsman's qualities within him, in order to bring the virtues of his tradition to bear in the new world. No one ever says it, but people who still tend to take responsibility at home always are chosen for such jobs. They are usually heads of families. They invest meticulous care in their work, not from manual dexterity alone, but from the worker's personal ties to custom and tradition.

I wish that industry would admit to the existence of this problem. They should send people with infectious attitudes — from logger to manager — and not factory directors with MBA's, to Persia and Afghanistan. I am pleading for a type of man who encompasses and combines the peasant's settledness and stolidity with the craftsman's reliability and resourcefulness in his work.

I am afraid that we have lost precious time in arguments, verbally splitting apart peasants, craftsmen, and workers. Europe has devoured itself by emphasizing their differences in the last decades. It is high time that we undertake a historical dove-tailing of the different epochs in the training of the working and producing man. We have to realize that we want to export the last 1500 years of European history, and not the momentary state-of-the-art achievement of some machine. I do know of course that the urgency of the moment directs our attention at first only to these deliveries. But looking at the position of Europe in the multiplicity of worlds with different backgrounds, we should affirm and support the unity of the European achievement which today should be used to make other economies unfold.

By the way, it is astonishing that the leaders — the monk in the early Middle Ages, the artist in the later Middle Ages, and the engineer in modern times — were all rather peculiar creative types. The monk had no power at all. The architect, the artist who created churches over one or two hundred years in either Florence or Konstanz, or Strassbourg, were not powerful men as measured by external raiment of power. But they had authority. If we turn from the peasant, craftsman, or worker to the modern engineer, to the former cathedral architect and to the monk who once settled in barren regions, we have before us the three groups of leaders whom we still need in modern society.

One example: the Americans sent highly-qualified engineers to Persia, all of whom failed dismally. They could not make themselves understood by the people there; they accomplished too much; and there was too great a gulf between their highly-refined training and the backward, several millenia-old economic system. The Americans were not prepared to overcome this gap in spite of all their generosity and goodwill.

You meet up with men in the developing countries who believe quite naively in modern technical progress, who start off perhaps making fun of the ignorance and dullness of their past, but who usually do not have any desire to distinguish themselves in the use of the new-fangled installations. Painstaking, tenacious work by the imported technicians from Europe may be necessary to complete a water main. If the natives are left to maintain it, it will be polluted the next day. There is no point in just dropping the technological stuff down there, only to discover upon returning months later that it is not being used anymore.

Here is the problem. One cannot just make men happy with something one thrusts at them from outside; instead one must plant it within them, so that they take possession of it for themselves; so that they develop their own relationship to it, and are proud when it functions. We face this task all over the world. The same things cause difficulties for the Christian mission. The Americans wanted to build a modern hospital for Albert Schweitzer. He rejected it with thanks, and said, "My people have to come into the hospital with their whole clan, or they lose their trust in me. If the father is sick, then the whole family is sick." The same thing, by the way, is true in the Phillipines. American doctors wanted to close the hospitals, because Filippinos arrived in groups of sixteen, the whole clan of a patient. This is a very serious thing. The white men usually do not manage to make a connection between the newest refined technology and the hearts and brains of these other peoples. I cannot understand why industry has not recognized this fact long ago and drawn the practical conclusions accordingly.

An economy of mankind is possible only when at least three generations step forward, when the living generation picks up those who have fallen from the tree of life, carries them back and grafts them onto the tree again. Do we believe that the Biblical words are just sayings? Every gardener knows that a tree must be grafted, pruned and fertilized. But we refuse to fertilize the tree of mankind. We take care of heads and hands but nothing else. That is not economy; it is waste and exploitation. The image of mankind as a tree is meant quite realistically. The historical novelty has occurred, that we have become one single human race, even if individual men still stubbornly cling to national or racial conceits. Within this new completeness of mankind, the states as structural elements move into the background and the overlapping powers of society determine our fate.

The society which is to produce this new economy must also supercede the mere concepts of "East" and "West", of the Soviet system and capitalism. Generations form a chain through time, and each living generation has the decisive function of performing its own important task. It does not matter what you call these tasks; they are not tied to programs which are of a passing nature. But we must remain able to recognize these social tasks under different names. I have been speaking of Mission of course, although I would take care not to call it that. In future society, certain tasks will come to exist under entirely new names, but they will still be part of the continuity of the gift which the monks gave to the peasants, the artists to the craftsmen, and in the age of technology, the gifts of the great natural scientists and inventors to the workers.

Nowadays when we have to synchronize all the work done on the planet, we will again have to train a group of men who set the standard for patient, slow and unobtrusive work without visible gain. Thus if the expanses of the continents of our planet are to be cultivated now as the Creator of the soil and its treasures demands, then industry, craft and peasantry must pool their gifts: the reliable dedication of the free peasant who need not be overseen, the reliable craftsmanship of a team, and the coherence of the antipodes brought about by enterprises spanning the globe. Where are we faced with this already?

The painter Hans Heysen, born in Hamburg, painted the Australian landscape. How did he come to it? The picture was used for the menus of the Australian airline "Qantas." The airline spells its name in a very un-European manner without a "u," and flies large jets; but otherwise everything about it is interchangeable with Lufthansa, or BOAC, or KLM, or SAS or PanAm. The Qantas stewardesses belong to the new human race we have been talking about. The staff of this airline from the antipodes deserves a new rank. I would call them planetary domestic servants, because their role is derived from the planet. They cannot fly more than 72 hours a month without falling apart. During those 72 hours spent flying around the globe they are not just in "some part of the world." The expression "world" would damage them because they are tightly harnessed into the service of the planet. That is their dignity. Their service is causing the planet to grow together. They suffer for its "one-ness" and mobility, and almost like astronauts, they have to give up the comfortable feeling of having a sense of locality and an old timer's close attachment to the soil. Their new virtues are harder and more painful than the virtues of the day laborer who is rooted in his home soil. Anyone who does not respond to their suffering with sympathy shows that he is outmoded and unfeeling. We owe this new race our future life. These members of the planetary household are making the greater social sacrifice, for they have lost their comfortable order of a single place and living in their own time. We are used to having reverence for the Benedictine calm. A Qantas team does not deserve less respect. But while modestly not asking for respect, they do demand something also of us: that we make place for these domestic servants. As they serve us, we should be ready to serve them in our turn. To do that we have to admit that we have paid too little attention to the shift-workers, the nightworkers, the telephone operators, and the security guards. But they are now becoming so numerous, so much the rule, and they appear before us so impressively that we will have to start adjusting ourselves to them. All of our celebrations are still local, they depend on the sun, the seasons, the climate, or the national history. The Church has tried ever since Christ to create a year of the soul which should both rise above local customs and apply to all times. But concessions to the countryside — which disguise a celebration of the soul with sun wheels, Easter eggs, Whitsun branches, or Christmas trees — can become pasted-up decorations too easily when a soul is flying through all seasons 72 hours every month.

We don't have a solution to offer, except that the reader should become willing to admit: the romantic qualities of the village and of hunting across peaks and valleys is an extreme which is increasingly confronted by the demands of a planetary household. The planetary household is too important to be treated as an embarrassing exception from the local custom. It demands equal treatment. Anyone who glorifies gardening, hunting, and fishing will be a buffoon if he does not wish to glorify the planetary household as well. The future is just as poetic as the past, despite romanticism. A new order will never be created unless we long for it. We are trying here to find the seed of peace between the future planetary household, and peasantry of the day-before-yesterday, and world trade of today. This peace will remain unthinkable as long as we romanticize the past and heave sighs over a future for which we long. Our breath should extend across many centuries. We should continue to inherit the attributes acquired by the peasants, the craftsmen, and the workers; we should retain our pride in the hermits, the architects, the engineers. We have to keep alive the attributes which others have acquired. Our planetary household's future life, its freedom, its historical character, its unity, and its coherence depend on this achievement.

6. The Planetary Household

Of course we think that our own situation nowadays, shortly before the year 2000 after the birth of Christ is overwhelmingly different from that of all the other bygone days. The heads, hearts, and kidneys of our forbears, however, were just as curious, sympathetic, and shocked as ours are. And although we have received a new assignment from the planetary household, part of it was dealt with long ago by "hospitality." The "una sancta," the ecumenical, the holy, the universal Church of mankind existed before there were popes, patriarchs, or UNESCO's. For good or for ill (and today, more often for ill than good), hospitality anticipated this final planetary peace, because the "hostis," the enemy, could become a guest. The same Latin word meant both, and made it possible for one to change into the other, just as a baby might become an old man; the young, old; or below, above.

The over-used dialectic has always been part of mankind, because you and I can assert ourselves only by exchanging roles forever. The dialectic of world history is that children become parents, brides mothers, and pupils teachers. This is no new discovery of ours. This old dialectic deserves more respect, however, than it gets. It means that each one of us will be called upon to play the opposite part some day. If an obstinate boy who contradicts his teacher will one day teach a boy just as obstinate, then the secret of dialectics on the planet is apparently very different from what little boys and full professors imagine. Actually, teachers and students, masters and apprentices, brides and mothers, voters and their representatives are all irrevelant per se. What is important instead is the ability to change at the right time from apprentice into master, patient into doctor, or subject into ruler. We must be able to call forth from within us the variation: teacher, or maybe policeman, or welder, or juryman, at the right hour. We hear and we speak because every day the same "homo insipiens" (Latin: unwise man) has to become "homo sapiens" (Latin: wise man). That is why we listen and we speak; so we are incessantly calling upon and renewing one another. Our world is the greatest chatterbox in the universe, because all changes and all ways of life in it have to be called into existence by mutual appeal and obedience.

To say that we are called into being is not just a pious expression. We can emphasize the differences between child and elder, between young woman and grandmother, between apprentice and master, only because they were previously called into being by all of us together. They are literally "counter-calls," contradictions. We most certainly don't propagate just physically. Speaking is also part of propagating our species. For a person who speaks, acts in the name of the human race. No word leaves the mouth without being spoken in the name of the human race. How else could the swindler cheat, the forger forge, the perjurer lie? All of them can mislead us only because we give their statements the weight of a truthful statement. A totally suspicious person couldn't live. We have to trust, even to cross the street. Someone once jovially asked, "Who is a pessimist?" and answered, "Anyone who looks both ways before crossing a one-way street." That's the extent to which we trust things which have been said. If, in fact, the street signs were liars, we would all be run over.

Now it is nearly unknown today that listening and speaking, that truth, and that calling-into-being are activities unique to our species, by which, from within our mute animal selves, the destiny of mankind and our own destiny within the human race are called into being. A person who speaks for our destiny always speaks for our species, just as every voter votes for the country. The Cartesian dream of thinking, adult individuals obscures this fact. Descartes (who was orphaned at an early age) dreamed them up. But, my honorable reader, I am sighing while writing these pages, for writing them is entirely against my lazy and hedonistic individual self-interest. I have to overcome my "self" in order to fill them. I keep telling myself that someone should finally proclaim this forgotten truth again. I keep encouraging myself, silencing all of my own objections to writing books. Whenever I become lazy, I reproach myself. It is the species which keeps urging me to call this book into being. And these voices contradicting Mr. Me and my resistance are actually the conditions for producing a worthwhile book. One part of me through which the whole species speaks has to overcome the other part of me, the selfish slob.

The poor individuals of the enlightenment became schizophrenic because they attributed speech to the "self." Conrad Ferdinand Meyer ended in mental derangement; earlier he had written, "I am not a cleverly-conceived book, I am a man with his own contradiction." It's a nice verse, but it endangers life, and it resembles sleep-walking, as does schizophrenic humanity.

"My contradiction" must be drawn out of myself, evoking contradiction in a soul which loves me. Only a person who is being contradicted by a fellow human being can have his contradictions cured. That was true for the thinker Descartes (who flirted with the Jesuit order but had children by a Calvinist) and it is as true for rich empires as for poor colonial peoples. It is true for modern China and for Russia and for India. Germany excluded every contradiction from within itself until it was destroyed, divided, and disproved. It exterminated the people of eternal contradiction, the people of Israel, and in a raging, self-determining lack of contradiction, it called forth its own downfall.

That is not as unrelated to the affluent society or the developing countries or the planetary household as it might seem at first glance. Mankind has always swayed between open contradiction and silent resistance. In healthy times there was contradiction. When sick, peoples have just resisted silently, and while obstinately resisting, succumbed to the illusion of being alone in the world. A person who is addicted to philosophies of the world (Weltanschauungen) will go crazy because fear of the world drives one crazy. Between 1871 and 1918, every German had a "world philosophy;" in fact even his own personal one. That doesn't work. As I have already pointed out, the expression, "world" doesn't really mean either "heaven and earth" on one hand, or "planet" on the other. The "world" has nothing to teach us — loveless and silent as it is — therefore everybody can form his own opinions about it. But you cannot use educated people's philosophic opinions to master either the "earth" or the "planet." The Earth demands that we work, toil and sweat. The planet demands that we participate, engage ourselves, and join up passionately. Your philosophy of life is worthless garbage compared to your sweat and tears. I'll take your philosophy seriously only if you risk your life for it. That is ancient wisdom. Living souls between heaven and Earth have never been able to afford the luxury of philosophic opinions. They have risen from being clods of earth to being members of the hosts of the Lord. The hosts of the Lord obey and believe without seeing and forming opinions.

For a long time such faithful souls have permitted the peoples to contradict one another. Only men who risk their lives have undertaken to contradict: the peoples have chosen two ways of letting themselves be contradicted, war and mission. But since the future of both wars and missions has become questionable, I am writing this book. I am only one of thousands who sense that a new day is dawning beyond the missions of the Church and the realms of the states. My excuse for writing one book more is that I have been searching for modern forms to replace mission and imperialism ever since 1912. I knew that I would have to look for them in my own back yard rather than in Africa or China.

Once during maneuvers as a sergeant of the reserve, I climbed up to the place where the Bonifacius oak grew, and met a hermit, a Franciscan monk, sitting in the open. It turned out that he had been the commander of a Berlin guard regiment. There he was now, sitting in the sun above the busy world, bearing witness to me (I also came from Berlin) of the peace of eternity above the busy paths of the world. The hermits have always carried out their mission by making time within themselves. "He had time because he made time," cries Ernst von Wildenbruch in "Edlen Blut," describing his hero. Yes, but what does it mean to make time within yourself? The person who could answer that would have the "open sesame" for the future of our planet.

Clock-time is ridiculous and was thought up by technicians.

But it always has to withdraw in the face of the truly serious inner time which two people in deadly danger offer to one another, and which allows them to enter into one common life. Someone who makes inner time starts a new time. Since the crucifixion of the Lord therefore, we have been entering continually upon a new, inner time, leaving the clock-time of the Bureau of Standards to the SS (Nazi elite corps) and similar persons. But anybody who presents us with healing time will become our saint.

A young man did that, a man who proclaimed and announced the times of planetary peace once and for all. He wrote:

> "I have to send you some bad news. I have been sentenced to death, I and Gustave G. We refused to enter the SS, then they sentenced us to death. You wrote me that I shouldn't go into the SS, and my comrade Gustave wouldn't go either. We both would rather die than cover our consciences with horrible deeds. I know what the SS has to do. Oh, my dear parents, as hard as it is for me and for you, forgive me if I have injured you, please forgive me, and pray for me. If I had fallen in battle with a guilty conscience, that too would have been sad for you. Many more parents will lose their children. Even many of the SS fall in battle. I thank you for all those good things which you have done for me since my childhood, forgive me, and pray for me."

This is William James' moral equivalent of war standing a test gloriously. This indispensible moral equivalent of war can be reoriented constantly by this young man's example. A man who accepts death in order not to commit evil deeds helps us to do the things which must be done to replace war, lest we lose our creative breath.

A friend of mine who works in a smelter in Holland read this letter to his wife and children on Christmas eve 1964, before reading the Christmas story according to Luke. The family broke out in tears, and the youngest said, "Was this young man Jesus?" That father added in his letter, "She understood without my having said it. We had a good Christmas holiday."

We make inner time by overcoming fear. Fear of the world yields to a deep peace on the planet if we make time. Without a victory over physical, natural, material, visible, measured time, the time-warrior would not be fit for service. Only someone who allows an unending amount of time will be able to replace the world-time of the airlines and the factories with planetary time. And that needs practice. But the hermit, the artist and the engineer have taught us exactly this. They replaced "hand to mouth" time with unending amounts of time. Cathedrals needed 100 years, the forests 70 to 80; and the trans-Siberian Railway was a project on which men worked from 1850 to 1917. In each case the economy stretched beyond one generation, even beyond two generations. And an event needs at least three generations, and demands two graves before becoming history.

The young man who did not enter the SS in 1944 has overcome death, thanks to the reader who has been deeply moved by him. We can find hope in that. However, since the military has unilaterally declared that death in formation is honorable and resistance to orders unthinkable, then planetary service will be very hard to implant. No one eager to carry out the most shameful orders should be admitted into the new service, even though he may not run away in the face of the enemy. This service needs as its founders people like the young man executed by the SS.

This discussion has yielded something very practical for the coming era. Peace must be articulated and explicitly concluded. Otherwise the cold war will continue. Peace is not a natural condition that you can fall back upon or return to. That is the basic mistake made by people who worship nature. No, war is declared, just as love has to be declared. Peace on the other hand, is concluded in spite of a grave, just as marriages have to be explicitly concluded. A declaration of war makes the conclusion of peace necessary. So a condition in which people don't speak to one another before a war is followed by an explicit declaration of war, which is in turn followed by the conclusion of peace. This is the miracle of speech. Serving the planet is not a tacit habit, and can never become mere routine, as you have to start doing it explicitly. It is preceded by a lack of peace, which has to be articulated and recognized. Peace can be concluded only after that has been done. So the service of peace needs a form, it needs to be articulated into existence and established by declaration.

People deny this obvious, simply inescapable truth nowadays.

You can smell the decadence when people use the idiotic phrase "aid to developing countries." After all, "development" means decay to anyone who is not just studying zoology. Anything which is left alone and allowed to develop will lack the love necessary to save it from this so-called development. You would need a mind limited to breeding rabbits to believe in development as a goal or task. "By itself, everything which grows, perishes, and what is growing deserves to perish." Development means decay, and South America or West Africa won't be able to avoid decadence as long as they ask the banks for interest-free loans.

The mere world will become a planet only when worldly development is stopped and replaced by loving participation. Anyone who thinks that this is just impractical hocus-pocus should consider the simple question: Will the South Chinese, the Congolese, or the Fiji-Islanders of the year 2033 order their turbines in Germany, will they allow their students to study in Germany and their brides to look for their trousseaus in Germany? We can claim that we are anticipating the coming hundred years in a practical manner only if we can answer these questions with a "yes."

You cannot "develop" participation to last for 66 years. Participating is nothing complex that needs to be developed, but you do need to sow a seed, to declare love, and to prove yourself. That will then bear fruit after two further generations; that is, after two funerals and the deaths of all those who are involved today.

That is why "aid to developing countries" is nothing but worldly wisdom, good for the next 24 hours. It won't last for a hundred years to come, because it tries to pass over our own long-suffering two-thousand-year-old history in silence. Such "aid" flaunts our progress, instead of praising such praiseworthy things as our time-honored patience, our lasting perseverance, and our faithful devotion. These are the powers that have to be exported in order to extract the poison fangs from mere technological development.

When economists make suggestions, it's never clear that they have studied economy. The noble word "economy" has little to do with figures, wages and prices. It means the household of the forces and motivations which surge through us and determine our paths in life. Because we are householders and embody an economy, we mortals are the passing bearers of eternal functions. We are like guards on duty, and have to be relieved.

Therefore the efforts of people serving the planet have to make themselves felt by the next two generations at least. Every soldier who falls in battle fighting for his country is sacrificing his life for the legacy of his forefathers so that his sons or descendants can build new lives once again on that legacy. So every war involves Tyrtaios' (the Spartan's) three generations, the old men, the young men, and the boys. In this context the words "aid to developing countries" sound a little funny. They ignore this chain of generations. Three generations can never be bound together by money, plans and books. That bond can come only from the faith of the elders, the love of the young, and the hope of both. If a warrior was unable to connect his forebears with his descendants, he remained a bully, a soldier of fortune, a mercenary. Men in future ages without war have to replace the warrior. But they must understand that he is a three-generational being. Without having some three-generational being, war couldn't be abolished because mankind would atrophy. We have to embody the whole race. As individuals, we have dignity only when we are embodying the whole race. A truthful, courageous, and believing declaration like the letter from the farmer's young son should give us the courage to realize that three generations truly belong together. I am the mediating generation and I am extricating him from death by burdening you, the reader, with him. One generation can indeed listen to another. I am transposing the dying boy into you, and he will endure through you.

There is no other method of enduring through time except this particular one, where the truth from the dying reaches those yet to be born through the living. So the services done for peace will have to last beyond the short span of life, spent in service at a particular time and will have to remain effective after the person leaves if these services are to be a moral equivalent of war. Seen in this light, some of the palpable achievements of the day necessarily seem unimportant, like musical accompaniment, or wrapping. Such wrapping may indeed be useful, but it will not be of much help in preserving a legacy for three generations. Wrapping like this dazzle us, and often enough makes us blind to long-term continuity. Daily fashions are only wrapping paper after all, although we presumtuously or over-politely call them "the spirit of the time," to cover up our cowardice and allow us to commit the evil deeds of our time, unpunished. Every swordbearer of the spirit, on the other hand, is a three-generational being who converts us by his sacrifices.

7. The Peace of the Pirates

A pirate is a tempter. That's all that the Greek root "*peira*" (Latin: "*pira*") contains. "*Peira*" means "attempt." People in Germany don't think warmly about pirates, except when listening to Brecht's *Three-Penny Opera* and *Gilbert and Sullivan*. They are landlubbers, which shapes their relationship to piracy. Despite Goethe's verse, "War, seafaring, and piracy are a trinity and inseparable," they have separated the three. German warriors have never recognized pirates as their equals. A postmaster or a militia lieutenant care as little for pirates as they do for highway robbers. The Lutherans at home found missionaries on the distant frontiers of the Church just as ominous. From Luther to the Pietists there was almost no mission in the German state church. There is a revealing joke about the various national characters: an Italian, a tenor; an Englishman, a missionary; a German, a scholar; a Frenchman, a hero. This is a registry of the personal attributes which the great nations have exported. The whole world loves the singer Caruso, the whole world reads the German scholars, but the whole world listens to the English and Yankee missionary. For that reason, my contention that a new type of piracy would be needed in the future doesn't affect English-speaking people, who are already brought up to respond like missionaries. However, in a country run by regional administrative officers, the name "missionary" just sounds petit bourgeois. Exactly what we should call our emissary of the planet may seem unimportant. The name "pirate" however has the advantage that it will annoy all Pharisees, all pensioners, and all petit bourgeois. Pirates even risk their pension rights! Horrid creatures! People don't want to hear about illegitimate children, Jesus of Nazareth notwithstanding, or about actions which go beyond the law.

The Germans' childish entrenchment behind an oath sworn to the beast from the abyss, Hitler, will be shown up for what it is to the children in confirmation classes outside Germany, but hardly to those in Germany. Ever since 1517 the Lutherans have celebrated the 30th of December as the day of King David, instead of celebrating it as the day of St. Thomas a Becket, the martyr for freedom. Thanks to this decision, ever since 1517, the law established by the prince, the Roman law, seemed sanctified, appearing side by side with the book of Leviticus. The famous lawyer Ernst Zitelmann was so horrified by this that he wrote, "That it is lawful is what's so very awful." It's not a lovely verse, but it is a true one. The boundaries of law in central Europe are so solidly established that it is impossible to imagine an area free of laws on the continent. An Italian Minister of Justice expressed the horror that many child-like people feel toward the legalistic man this way: "What kind of country can Germany be, where they obey all the laws?" The word "anarchist" is simply an insult, and neither constitutional law, nor ethics, nor catechism, nor psychology, nor even the poets have saved us anywhere a piece of God's world that is free from the authorities. Chaos, cried Friedrich Nietzsche, who was revolted by all these legal paragraphs, should give birth to a dancing star. But the poor man was considered crazy and became crazy. The word "chaos" as well is a mere word used by people with a refined education. And in front of well-behaved school children, "chaos" really cannot be made believable or attractive. That's why I'm going to try the word "pirate."

Piracy is the attempt to assert oneself in the absence of the authorities. And because that seems almost unthinkable on dry land, when you hear the word "pirate" you think of the high seas. Innumerable times, however, islands or harbors have been occupied because of piracy. So pirates who remove themselves from the civilized world are just extensions of Robinson Crusoe.

Since the pirate is not counted among the recognized authorities, he is hard to define, except in a negative manner. In logic this is called "per exclusionem." Piracy doesn't follow the laws of recognized states, but apart from that, piracy includes a range of widely divergent acts. First, they are morally different from one another. The most noble actions may be counted among the deeds of pirates: rescuing shipwrecked persons, freeing slaves, protecting women, preserving valuable documents, paying ransoms, feeding the hungry, or extinguishing flames at the danger of one's own life. The standard text book on international law of Roland R. Foulke (Philadelphia 1911) correctly says:

> "The idea of acts of violence or crime, which are usually characteristic of piracy, are not essential ingredients. A body of pleasure-seekers, without the slightest idea of harming anybody, who should set sail, haul down the flag of the state of which they were formerly members, and hoist a flag of their own, could be seized and taken into port by the first warship of any independent state they might fall in with, and the vessel condemned without hope of any restitution. They would be pirates."

It would be better to think of Paul and Barnabas if you want to appreciate the spirit of pirates.

In the early days of Christianity at the time of the Emperor Nero, the Christian missionaries seemed like pirates on land. In those days there were still holes in the highway system of the civilized states. And these highway-robbing Christians used to steal the income from the local gods, just as the Corsairs or the pirates levied contributions from the harbors. Obviously the day is rapidly approaching when the seas will be as completely given away as all the continents of the planet already are. And the freedom of travel through the air above us will soon disappear for unofficial balloons. I myself was once able to fly in a balloon, and cross over national borders without being announced. When earth, water and air have all been given away, however, the pirate, the anarchist, and the missionary will seem to us like dinosaurs from prehistoric eras, for the idea that there could be a "natural space" free from human regulation will have disappeared.

This book has been written for that moment. Once the policemen impose the same borders as the geographers, the child of God within us will suffocate. The double pressures of the law and of knowledge will overwhelm our consciences and make us believe that no one can fly, write, love, hate, or serve except on the legitimate paths of the railroad, the airlines, the ski-lifts, the various academic scholarships, the state exams, the draft, the street signs, the IQ tests, and birth control. Nowadays already, each new-born child in the state of Arkansas receives a serial number. So we all may be considered "seen through." The day when this madness is victorious, books will stop making sense, the Bible will stop being understood, the free originators of any kind of future will be considered crazy and will be locked in the institutions in which crazy people used to languish. It has already nearly come to pass that anyone calling for a peace which surpasses all this socialist reason is considered crazy.

I have the honor to have been considered a public danger more than once in my life. The first time was in 1912, when I wrote, "Language is wiser than the person who speaks it." My thesis almost foundered on this disturbing reality of the Holy Spirit which I had perceived. Balaam's ass was considered unscientific! It occurred a second time in February 1919 before the Versailles Treaty was signed. I called upon the German legal profession to offer their protection to the Kaiser and the generals, and let the Allies — as they in fact did in 1945 — put the legal profession on trial instead, so that the indispensable new planetary law could come into being, once we had been made liable. The reader can imagine the effect. I wanted to do away with the immunity of the German professor! So I left the "institution of higher learning and limited liability." It happened a third time in 1935 in the U.S.A.: I was considered a blemish on pure science, because I referred to the workings of the living God in a lecture hall. Again, Harvard University, objective, free of "values," felt that it could restore its honor only by contemptuously shoving me off into the theology department. These three opportunities were related to our three articles of faith, 1912 to the third, 1919 to the second, and 1935 to the first. I was so audacious as to invoke them. Presumably I have survived, thanks to this trinity. It cost me my native country, my colleagues, and so-called science.

But in recent years people who have had the misfortune to have to call upon all articles of faith at the same time instead of just learning them by heart when children — people like the Moltkes, Haeftens, Delps, Bonhoeffers, Kleists, Schwerins, Lebers, and Reichweins (all active anti-Nazis), have had to give up their lives.

The distorted thing about our position today is that almost nobody who is earning his daily bread with the Christian creed realizes that the credo may be deadly dangerous. Only laymen believe that, or people like Bonhoeffer who take off their frocks. If the creed is not considered dangerous, divine worship is emasculated. The creed is either high voltage or empty straw.

As a result of this dilema, I cannot help placing the future planetary service under the pirate's flag. How else could you understand me? Piracy is after all the last bit of freedom from regulation, whose purpose our fellow men understand, and which we are still considered capable of committing. That is indeed indispensable, for we need a sphere for our planetary service which no government or governing body takes care of or has to be responsible for. The accepted textbook on international law says as eloquently as soberly, "All pirates have this in common, they make it impossible to hold any state or church responsible for their actions." What great people they are! It is said of them, that they understand the knack of making it perfectly clear that no one need feel responsible for them. That requires great skill.

In the age of statistics, this skill is more important than any other. For anything that doesn't occur in statistics is considered either non-existent or at least ineffective. One of my best friends assured me at my own table: "You are statistically unimportant." I couldn't care less. The TV-drunk public, however, and the professors of economics deduce that the statistically-unimportant doesn't count, and they draft their laws accordingly. I once gave a learned and thoroughly-researched lecture on the old saying, "Quod non est in actis, non est in mundo," "Anything not on file doesn't exist." Nowadays that saying should read, "Anything not in statistics has no role to play."

Now it is most odd that Luther, like the modern pirates, was trying to upend these two sentences: "You are statistically unimportant," and "Anything not on file doesn't exist," with his much-overused expression "By faith alone" (sola fide). What Luther actually meant was that only things not on file and things which cannot be registered statistically were useful for peace on Earth and good will toward men. So the pirate's peace is really no more than a lay translation of the reformer's sola fide. His expression has been sucked dry, unfortunately. The expression, "the pirate's peace" might still be forceful, because at least the pirate doesn't belong to any state. He is not a member of a visibly-organized system of law or of a diplomatically-recognized power. His actions are unique enough to prove that he has the intention and the mandate to shake off or even explicitly repudiate any authority or any power which at the first glance seemed to rule him or to which we might have to ascribe his actions.

If the reader were to ask me what pagan antiquity called those unregulated pieces of the human world, the answer would have to be a word that is very similar to our soldier of fortune: a person who hires himself out hither, thither and yon; one who occasionally wanders alone through the countryside as a brigand or a highway robber, like Schiller's "Robbers." Such a band was not explicitly internationally-recognized by anyone, and was called "latrocinium," from the words for "soldier's pay" and "mercenary," without declaring their deeds good or ill. The thieves on the cross were called "latrones" as well. That is never explained in the clerical wisecracks about those poor devils. This word for these thieves is basically not a derogatory one. It doesn't impute ill deeds to the person, as "thief," "murderer," or "criminal" would. It is derived from the word for daily wages, "latris". So in ancient days, anyone who worked for daily wages might be called a "latro." This "robber" lacks a lasting and formal recognition by the other groups of mankind. He lacks what American sociology loves to call a "status symbol" nowadays (another overused expression). American sociologists claim that every one of us would sacrifice money and honor for these status symbols, be they Cadillacs or an invitation to the White House.

Now pirates at least don't make such sacrifices, any more than the publican in the Gospel. The pirate renounces status symbols! Since such a thing is considered impossible in the modern world of atlases and encyclopedias and "Who's Whos", I have to ask the reader to imagine a future society which ridicules the statistically-filed and alphabetized and computed world of the electronic brains, and which intends to establish itself upon incalculable, unknown human entries who have not yet been named or recognized; a society not presided over by a bank president, but by the child blessed by the kings of the Orient. Anyone who wishes well for the future beyond the world of today will learn something important from the pirates. The name pirate, given them by inimical nations and their lawyers, totally ignores what the pirates actually do. A pirate can be virtue itself. One doesn't become a pirate by committing a crime, but by being refused recognition, for good or for ill.

It is odd, but now at the end of "world" history, when there are nearly no pirates left, the same urge which the pirates have has become the desire of all the nations. Nations want to base their own status on their own self-images, just as pirates do. The existentialists have to "understand themselves;" the states have to direct their own fates; the artists have to explain themselves; and the scientists have to describe themselves and the whole of science these days. A healthy people, a healthy family, or a healthy child of man would let himself be called into life, appointed, designated, and directed by his neighbor. This self-conscious independence of all worldly spirit does resemble one-half of piracy. Pirates also forego waiting for or hoping for recognition by society. But the nations and the individuals make into their very own driving force a situation which pirates accept only of necessity. The nations boast about it: "My authority to recognize and know myself is all that's needed. I don't have to be recognized by anybody else." This proletarian attitude — "We of God's wrath are the proletariat" — seems to describe individual nations today. The peoples did not behave like that before 1870.

The United States literally began its existence with the statement, "A proper respect for the judgment of mankind." All healthy nations have spoken with one another and received their life's breath that way. Speaking with one another is a proof of spiritual health. As long as we listen and respond when spoken to, we can build a "co-responding" relationship. Ever since 1870, however, the soft-headed European nations have been raving about national myths. Whom should they "co-respond" to? The Devil whispered, "to yourself!" And indeed they told themselves who they were. They proudly cultivated their own myths; to put it simply, they lied to themselves. They forgot that myth is a boomerang, something which a madman would use to appoint himself Emperor of China, for which, however, he would then have to be certified mentally ill.

Telling yourself who you are is the most common spiritual illness among both individuals and nations. To be at peace is to know that others allow us to give them their names, and that we allow others to give us our names. Reciprocal naming, "co-responding," is the way you conclude the Biblical peace which surpasses all reason. It surpasses reason since it always reaches beyond comprehension. Simply because when we speak to one another we enter God's "una sancta" which the myths deny. Peace actually means, "You are not really alone, and you can tell someone loves you when you hear your name in your good Samaritan's mouth." When a person hears his name from his neighbors and fellow men, he can be assured of their brotherly conduct.

When my wife and I were put in a position which compelled us to carry our double name, Rosenstock-Huessy, our parents went before the judge and these five people who were so important to us were the first to call us by our new name. They put us on the right path. Healthy people ask one another the time of day, and greet one another by name. Because my acquaintances on the street address me by my last name, and members of my household address me by my first name, my name is what they call me. Christ's name is Christ because we recognize the child Jesus of Nazareth as our Christ. All of Christianity rests on that foundation alone. It says so explicitly in the New Testament.

Pirates can show you how to be free from the authorities. Pirates, who received their character from the non-recognition of the surrounding world, don't tell others who or what they are. They accept the curse of supposedly belonging nowhere, and leave it at that. Pirates don't sing "Rule, Britannia" so we can trust them.

I am talking about pirates because I couldn't indicate the entrance into the coming era otherwise. The inhabitants of the continents oriented themselves by watching the heavens. The explorers of the world have discovered how to fly. The messengers of the planet will have to grab the globe away from the raving nations.

Wanting to determine its own fate empties a nation. Lacking the justice which accords each its own place, a nation becomes, in the the words of Augustine, a band of robbers. If the states themselves are becoming similar to pirates, I will have to seek shelter with the pirates myself in order to show my readers the way to the source of health. Let's look at the quaint phrase of international law once again. "All pirates have this in common, they make it impossible to hold any state or church responsible for their actions." This formulation is as original as it is unexpected, for in our classifying and classified world of encyclopedias, registers and files, pirates know the art of slipping through every list or catalogue. Hats off to them! This behavior is unheard of in the age of statistics. Here are people who act first and then seal their actions: "We can't be organized, summed up, or registered anywhere." That would be an impossible accomplishment for philosophers, statisticians, or natural scientists.

Anything which can not be catalogued is considered either nonexistent, or unimportant and ineffective. "The German aid to developing nations amounts to X million marks, which proves that the lousy 96 volunteers in the British Peace Corps are statistically unimportant." That is an unimportant sentence for serious men, as the entirety of statistics disappears in the face of one single important name. There are two or three billion women in the world, including a certain number who are of marrying age. But language seems to stick its tongue out at statistics, as the expression used in the real biography of John Smith is: he goes and marries Josephine Jones "of all people." Yes, "of all people." He is actually stopping his marriage from being precalculated by any population statistics.

In spite of that, the obliging reader won't mind if I remind him of some truths which he has already experienced. We aren't reasonable people when reading, thinking, discussing, giving lectures, or watching television. A third figure emerges inside us, a part of us, in between bachelor and bridegroom, daughter of the house and lady doctor. This in-between figure listens to statistical truths with devoted trust. This third figure in you and me is the part of us which is a member of the listening public. When this public aspect takes over, we change from a person into a curious man, from a husband into a listener, from a music-making amateur into an enthusiastically clapping member of the audience at Bayreuth. But what is this part of us, which is a member of the listening public actually? It is the unfinished part of us. We are members of the listening public today, to awake tomorrow as members of the ranks of educated men. We are the listening public today so that we may fill ourselves with illustrious examples for life's decisions tomorrow. "Everyone must select his own hero, whom he can follow to Olympus." Being a member of the public exposes us to the choices that we will have, in order that we may vote wisely when the time comes. Nowadays we are inundated by the mass media (as they are called by the Americans) to an unprecedented degree. That's why city man who is incessantly exposed to them also tries to keep using their language when he makes his final choices, although for him that language is meaningful only as long as he is a member of the public.

From afar, statistics, ideas, and psychology are completely harmless. Close up, however, they might possibly mislead you to the revolting conclusion, "Everyone looks out for himself first." Before you beame a member of the listening public, you wouldn't have even understood this shameful sentence. And once you have encountered serious life, you can see that a lower form of man who has remained stuck as a member of the public is a human toad grinning back at you from this arrogant and silly sentence. As members of the listening public, we are in an irresponsible in-between condition which inundates us completely and is with us after childhood and before we enter serious adult life.

Because this giant public exists, I have to exempt our role as pirates explicitly from all ciphers and figures and concepts and definitions. I could never make this role comprehensible to the public figure within you. One of my best students covered the Russo-Finnish war as a reporter in 1939. After experiencing the defeat or rather the capitulation of the Finns on Good Friday, and being drunk with his own rhetoric, he wrote, "I have seen Christ on the cross there," and he went on to Sweden, and to Belgium continuing as a war correspondent. When he came back I put it to him that he had blasphemed, and he knew it himself then. For no one "sees" Christ, unless he stops being a reporter and participates in

Christ's suffering. We can live as "listening" public while preparing for our own drama, for although we are indeed sitting in a theater, it is in fact an antechamber of the room in which we will have to give our own answer.

I would not insist so pedantically that the reader set aside his membership in the public, if it hadn't become a real perversion nowadays. I recently read, "The love of women prepares us for the enjoyment of art." Turn it about and it will be true. True art prepares us for love. Art serves love. It enhances love. So I must beg the indulgence of the reasonable reader. I have to keep the peace of the pirates separate from the nosy public, from the intellectually-curious psychologists, and from the number-hungry statisticians. You should reflect on your duties and the trials they impose in the rough and tumble of everyday life, not while relaxing with a newspaper or in front of a television set. Then you can make decisions.

Here are several conflicting examples of that rough-and-tumbling modern life.

John Kosa, a director in the medical research division of the Harvard Medical School, wrote in the *New York Times* on January 20, 1965:

> "The weaponry expert of our times, who is active in so many countries is able with the aid of his technological knowledge to command a tremendous destructive power which can be at any time turned against the whole or a selected part of mankind. As long as he carries out his activities in his home country, we may assume that he is governed by the same moral considerations as the public life and politics of his home country.
>
> His decision to develop and use destructive weaponry is somehow subjected to the same forces of moral responsibility which very imperfectly but still to some extent control international politics.
>
> The wandering weaponry expert, however, is not bound by a sincere allegiance to one country or the morality of international politics, and one may wonder whether he is moved by any voice of conscience other than his self-interest. Evidently he wanders from country to country offering his skill in producing destructive weaponry to the highest bidder.
>
> He represents the greatest danger for an irresponsible use of modern weaponry because there is no guarantee that he will not sell his lethal skill to a power inimical to our life, our value, and our civilization.

It is in this context that one must consider the wandering German rocketry experts whose presence at one time or another was reported not only in the United Arab Republic but also in Franco's Spain, Peron's Argentina, and in at least one other country under a dictatorial regime. Their case has nothing to do with whether one is personally involved in the cold and sometimes not so cold war between Israel and the United Arab Republic. This itinerant group presents a general problem that sooner or later we must face and solve: the activities of the wandering weaponry experts cannot be left uncontrolled; they must either be stopped or brought under international control."

The man from Boston wrote all of this. A bomb expert who enlarges Hiroshima and delivers the weapon to the highest bidder is concentrating the secrets of the world at a particular point. He is exercising his power within a world order that no longer applies to him: he can do as he pleases, thanks to the old trinity of "world" law: freedom of travel, freedom of knowledge, and freedom of contact. We have carefully distinguished between "Earth" and "heaven"; "world" and "planet." So the reader will easily recognize the bomb-smith as the crowning point of the "world" history of war, which started with the First Crusade and ended in the present day. For the "world" is the part of creation which has been made freely available by the uniting of human minds. Prometheus has turned the earth on which we find ourselves into the material called the natural world. By comprehending it he can enjoy it, use it, turn it into dust and ashes, turn it into the Tower of Babel, or into St. Peter's, for that matter. Turning matter into the bomb is the highest triumph of a four-thousand year-old attempt to comprehend the world.

So John Kosa isn't shocked by the bomb itself, but by the fact that we aren't all working to comprehend the world together, that any individual could manufacture the bomb for himself alone. Such a highwayman, mercenary, or scientist of fortune gone wild could claim our commonly-won know-how as his own and use it.

That's the way with things of this world. Everyone fancies that he can conquer mere nature for himself, and he laughs at those who come too late. All that matters is who first found the gold or the petroleum, or the electricity, or in our case, the bomb. He who comes first, bombs first.

There is no escape within the "world." Since this word actually means the "world outside," we are either hammer, or anvil, as Goethe says in his poem. If we lived only in the world and on the world, then we could answer Dr. Kosa with a shrug of the shoulders: "Poor dreamer." But on the same day that Dr. Kosa's letter appeared in the *New York Times*, the following item appeared in our local paper. The whole state has only 384,000 (now 440,000) inhabitants:

> "Robert Maxfield left here Saturday to join a group going to Peru on a program for the Peace Corps called Rural Community Action and Agriculture.
>
> Maxfield's home is the Lemax Farm operated here by the Maxfield family. He graduated from Hartford High and later studied radio and TV repair.
>
> While in the army, he was stationed in Germany and visited France, Holland, Norway, Sweden and Denmark.
>
> In 1957 he spent six months in the International Farm Youth Exchange in Ecuador. There he took pictures and gave illustrated lectures on his return to this country. He has attended every IFYE alumni conference since.
>
> He took his training for the Peace Corps at the University of Oklahoma from Sept. 25 to Dec. 19. He already spoke Spanish but studied it again in a course that was designed to make the young people familiar with colloquial Spanish in rural Peru.
>
> They received health and sanitation instruction, and also had courses in the historical, political, economic and cultural aspects of the area.
>
> They went on a four day camping trip in the Wichita mountains for physical education and recreation. Here they took part in rock climbing, rappelling, belaying, trekking and horseback riding. They also were trained in self-defense and swimming.
>
> They spent four weeks in the Toluca Valley in Mexico on field work. In this area the conditions are similar to the ones in Peru where the group will work. Here they studied community development. Maxfield assisted in the construction of a school at Jiquipilco and observed the methods used by the Mazahua Indians in growing corn. They lived with local families and ate native foods.
>
> Locally Maxfield studied the process of cement block making to be adapted to primitive methods and the construction of ox yokes in the hope he can introduce them to replace the head yoke in Peru."

President Kennedy founded the Peace Corps in 1960. Even its name hints at the fact that it was founded with foreign countries in mind, and not for service at home, in contrast to the Civilian Conservation Corps founded in 1933. After only four years, however, the pendulum had swung back in the direction of service inside the country toward a "Civilian Conservation Corps." R. Sargent Shriver, the leader of the Peace Corps, has undertaken to build a service for the inner-American areas of poverty as well. Over-hasty critics have attacked Shriver's double office. But both branches of a peace service corps, one in one's own country, and one abroad, unfailingly call one another into being and imply one another.

The reader has already become acquainted with the thorough preparation of the Vermont farmer. Such a long-starting period isn't always possible. Here is another example, which is more like a dive. A girl wrote:

> "This is the hardest thing I've ever done. Absolutely nothing is familiar and I often feel totally alone — the physical difficulties actually help, as they take my mind off myself and the feeling of suddenly being cut off from the rest of the world. You cannot imagine the gulf between East and West, and it makes me laugh now to think that I expected to bridge it with a smile and a handshake."
>
> — Patricia MacDermot
> — Third Annual Peace Corps Report, p. 47

Patricia MacDermot, 24, who grew up in a big city, wrote these words three years ago, in the first year of the Peace Corps, when she was a teacher in an isolated Philippine village. Her words, which she was fortunately able to articulate, remain today a good definition of the "culture shock" which puts every volunteer to the test: can you make it, or can't you?

Bud Weisbert taught English in the Thai city of Suphanburi. After his first year there had passed, he wrote:

> "If I've learned anything here, I've learned I know nothing of the Thai people. What I have learned can only be expressed as my relationship with Mali, Kamone, Pi, Ubone, Achan Pin and my other friends and acquaintances. The unique quality of having a home in Thailand is the individuality of the experience from reading about Thailand or talking to others about Thailand. . .
>
> And to get to my number-one concern: that which I need to understand most, there's me. It would be impossible to estimate how much I have changed, how much I have learned, what I now easily accept as part of the world, and what I now can or cannot understand. Change has, of course, taken place, but I am too far removed from the

old to measure the distance to the new. . . I have learned a new way of speaking about things and a new way of looking at things. My ideas on love, life, death, religion, etc., have met ideas that grew up in a different culture — I have not yet met any almost-middle-class, sometimes Jewish, Thais who came from West Los Angeles.

And of my friends' relationship with me: most of them have learned a bit, too. They have learned to replace their stereotyped picture of what an American is with their feeling towards me as an individual. At the same time they realize that I am as different from all other Americans as I am from each of them. . . In Kanachanaburi . . . there is a cemetery for Allied soldiers who died while building the bridge over the Kwai River. On one of the headstones the following is written: 'Into the mosaic of victory this precious piece was placed.' I feel that we here are doing a job and placing pieces into the mosaic of mutual understanding, and that understanding in turn will occupy a place in a mosaic of the peaceful world."

The Volunteers themselves make a humorous distinction between the "Albert Schweitzer complex" and "culture shock." A 22 year-old teacher in a Peruvian slum — Tom Carter from Oregon — expressed this very well:

"I get a lot of letters from people saying 'how exciting your work must be' or 'how picturesque' . . . Volunteers call that the Albert Schweitzer complex. . . . I live in a picturesque bamboo mat house I built myself. I buy my water from a picturesque boy with a burro loaded down with water cans. I read and write under a kerosene lantern, sleep on a cot, and cook on a camp stove. There comes a day when all this suddenly becomes no longer picturesque, no longer quaint but furiously frustrating and you want like crazy to just get out of there, to go home. This is called "culture shock." It happens to one and all, usually about the third or fourth month. . . . I don't think I'll sign up for another stretch but you can't drag me away from this one."

At home Elizabeth Chalmers had worked in a cancer research lab as a technician and had taught biology. After two years in Pakistan she wrote:

"When you arrive for Peace Corps training with Education in the top left hand corner of your credentials, don't assume that you are going to be a teacher. Two weeks later you'll find yourself training with the laboratory technicians. When you arrive in a country to work for two years as a laboratory technician, don't expect the situation to be permanent. Ten months later you'll find yourself a teacher . . . for such is the challenge of the Peace Corps. . .

In September I found a school that was overcrowded and

understaffed . . . I joined the staff as a science, English, and physical education teacher . . .Shortly after my arrival the school closed for a month's holiday. It is fortunate that within the structure of a teaching position, there are many things to do when your school is closed as often as it is open. On observing the library in the teachers' common room, I casually commented, "What a shame the books aren't used more often." It took me two weeks to clarify their misinterpretation of my idle remark because . . . they were ashamed. With the aid of the teacher librarian, we set about to rectify the situation . . . We removed the books from the hands of some of their most avid readers, the bugs. After spraying and dusting and partly rebinding we began to record the books . . . and started a card system library. We located a small room . . elected student library aides, and opened a diminutive Dewey Decimal System library.

The new science building was near completion when I arrived and I looked forward to helping my co-worker set up a laboratory; but soon I found that construction had been stopped for several months, and no one knew the reason. I started my new job as construction engineer, campaigning for the completion of the building. I anxiously awaited the results of my campaign and was delighted when I glanced up from judging a sports competition, to find workmen painting the building. A small achievement but a tremendous satisfaction . . . For the past two years I have been a laboratory technician, teacher, librarian, and construction engineer. I have been a day laborer, piling dirt on a project doomed from the beginning to failure, a status symbol, and a friend; but most of all I have been a crazy American willing to attempt a project I knew little about, because it ought to be done. I would personally consider some of the work which I have attempted here a failure, but for the fact that some of the efforts may be continued. It is enough to be flexible, to try and fail and try again; in each attempt there is a small measure of success. There is now a brand new science building completed because I tried to be a construction engineer. It doesn't have any electricity, or water or furniture, but a building is completed whose construction had been stopped for over a year . . . I have learned to relax and accept the fact that eventually the things will take hold which I helped to start. One doesn't need to know more."

One doesn't need to know more. The chapter ended with that sentence. That sentence can replace the listening public's dogma of wanting to know about everything from contraception to the Virgin Birth. A serious bride of life would say, "one doesn't need to know more." That sounds very taciturn, but the sentence is rich in content. I think that the books of the last ever-so-curious centuries have not worked their way through all the unending variations on this sentence.

With this sentence, the girl in the Peace Corps has introduced a whole school of social thought, a "not yet" school. The "not yet" will have to join battle with the listening public nowadays. We admitted above that we can switch on our public aspect which watches life curiously as it rushes by, that figure between the old Adam within us and the rebirth of the decision-making free man. We are members of the public part in this immense interval: after hours, in the movies, in the theater, when we are just waiting for real events. A Peace Corpsman, however, in quietly saying, "one doesn't need to know more," is not acting as a member of the public. She is no doer either; she is jumping aside, and is not unlike an observer standing on the sidelines. They do look confusingly alike. That's why modern society doesn't have a name for this "not yet" attitude. Now, teachers, priests and parents sow seeds into those entrusted to their care. How does such a sower live after his seed is in the earth? Is he as unconcerned as the viewing public which just watches? Is he as concerned as he was then, when he decided to sow his seeds or to go to Nepal or the Himalayas as a Peace Corps volunteer?

It is almost unbelievable that we should use pale, nondescript expressions for the sower's time of expectation. I can hardly make myself understood to my readers, when I maintain that the whole future of our planet depends on making this "not yet" (Latin: non iam) time to bear fruit. We need to learn to live in that strange time after our will and skill have already done all they could, in which we await the result. "Who is such a person, actually?" I am asking. How should he live during this time of waiting? Such a person obviously can't stand proudly to one side, crying, "Look, I've sown my seed here!" Nor can he forget that he put his whole heart into this treasure. Yes, but what has happened to him? The sower acted in sowing his seed and he cannot forget it. Even so he'll ruin it if he takes the credit for it. That is the true dilemma: ever since the Lord, the greatest sower in human history, cried on the cross, "Oh! my God, my God, why have you forsaken me?" many sowers have gotten too much credit, and many too little.

What about our brave woman soldier who said, "one doesn't need to know more"? Who will remember the two-year service of those men and women? Who will admit that she was the one who took the most necessary first step, that first step which may reach its goal after fifteen years, perhaps?

Anyone who cannot separate boldly the "time" created by technology from the time of instinctive loyalty to one's profession will associate service for peace with mere service in uniform, or the technology of the experts: but the moral equivalent of war would never come into being then. It would be made to resemble garrison duty, or target practice, or at its best, the Army's peace-time maneuvers. And that would kill it. Just as war is too serious to leave it to the generals, so the Peace Corps is too serious to leave it to the experts.

Future society is threatened with the fate of metamorphosing, on the Doomsday of history, into a heap of seconds and days from an engagement calendar. Opposing this threat is the weak, and unpresupposing person who has stayed sufficiently free from engagement calendars to sigh, "One doesn't need to know more." Two incompatible types of time are colliding here. Church officials only use their engagement calendars, which know and plan everything in advance. The child of the planet makes bold to live in a different time, in a time of unknowing "not yet." Is it possible that both times should exist side by side, that both have to exist? That would deserve careful proof.

Indeed, I maintain: The coming generations will have to master two completely different types of times. And the rest of this book should prepare the ground for a peace service corps which has nothing to do with the churches and states of the last thousand years. It must be explicitly protected against getting confused with military garrison duty or a work service in a patriarchal, pretechnical village! Both threaten to devalue it. The bureaucrats and soldiers would like nothing better than for this undesirable "foreign aid" to be devalued that way. Oh! it certainly could be devalued that way! Love can be prostituted, war can become murder, the service of free souls could sink to the level of being a mere assignment. In 80 cases out of a hundred, the service may indeed be prostituted. I want to break a lance for the other twenty true servants of the peace by surgically removing everything the reader learned in school about state and church, seed and sower. It was all learned in an era which has expired.

The so-called time of the technological age is robbing us of our own real time. And anybody who sees technology descending on Alaska, on the Zambesi, and on the Yangtze Kiang, knows in his heart that it is high time to beat back the raging time of technology with the time of healthy hearts. "One doesn't need to know more" is what that blessed volunteer said about the fruit of her activities. But we are allowed to know more about the seed which she represents herself. What if the person serving were the seed that needed to be spoken about, to be sown, to be protected from the weeds of church and state? What if we were to teach people about both kinds of time? Technological time and fruitful time?

In 1739 an old English minister wrote about what happened during the election of his successor: "I wish, as much as you, that the affair might be speedily issued; but you know that things of this nature, in which many, and those of a different temper, are concerned, must proceed with all tenderness and voluntary freedom, without the least shadow of violence or imaginary hurry. Men love to act for themselves, and with spontaneity; and as I have sometimes observed, have come at length cheerfully and voluntarily into measures, which they would have opposed, if they had imagined they were to be driven into them."

8. David Scott Palmer

A news item of February 19, 1965:

"The police in Caracas, the Venezuelan capital, shot at two American Peace Corpsmen. One of them, 24 year-old Joseph R. Rupley, died immediately, the other was severely wounded."

There were nearly 250 Peace Corpsmen in Venezuela at that time. The country is close to Communist Cuba, so the Venezuelan political climate is highly explosive. Rupley's father — the family lives in California — responded to the news of the death with remarkable self-control: "We in the family are shocked and horrified by the tragedy, but beyond that we are aware of the fact that Bob was proud of his Peace Corps, and proud of what he was doing for our country."

The native Jeep driver who drove the two North Americans into their misfortune thought that he was being attacked by Communist rebels. In reality his Jeep was halted by the police who thought that there were Communist rebels inside. Although the young men climbed out of the car with their hands up, that is to say, defenseless, they were shot down by the policemen who were terrified themselves. Another Peace Corpsman — most skeptical of the policemen's motives — said to me with a sigh, "How trigger-happy such men of the law can be!"

The cruelty of white terror is one of the incomprehensible truths about civil wars, whether they are being fought in Savoy in 1815, in France after the Edict of Nantes, or in Germany after 1933 or 1525. It is understandable, however, that the cruelty and senselessness have been increased by modern scientific methods. And because the future planetary service is standing in the spotlight of this scientific method, I would like to elucidate this increase in cruelty with an example.

High in the Cordillian mountains of Peru there is a university town called Huamanga, which is connected with the capital, Lima, only by air. A great Peruvian teacher made it his life's work to turn this school in Ayacucho into the spiritual center of the distant province by combining all educational levels: simple instruction, training and the eternal teachings. That was 15 years before four young Americans, one of them a young woman, arrived there as Peace Corpsmen. Cuban sabotage soon undermined the Yankees' activity, and after a long struggle the four were chased away. They were treated badly, like the poet Cinna in Shakespeare's *Julius Caesar*, but I am mentioning them here because of two accompanying circumstances which cast a lurid light on our times.

First, when the Communists had forced the withdrawal of these "emissaries of American imperialism," the leader of the Communist band came to one of the North Americans and offered to hire him, saying they needed him urgently. They were fighting not him but his government.

Second, the Peace Corpsman so honored said "no, thank you" and returned to the United States where he wrote the story which is being printed now. But how did he write it? In the proper scientific manner, without mentioning any personal names. Even the hero of Huamanga, the university reformer, is referred to without his name, just a "Peruvian." That is called science nowadays. "How one coughs and spits, is something learned from physics" (Goethe). The other figures in his battles are even more overcast by the shadow of his scientific boredom, pardon me, of his abstraction. As a result all the actors disappear behind the labels "Peruvians" or "Yankees." Peruvians are one species, Gringos another. Of course they are mutually irreconcilable as types.

Max Weber did away with Christianity in the simplest way: just by calling the one and only original man in human history a charismatic figure. Science castrates all effective men by ignoring their names. These unhappy castrati have no choice but to sink to the level of pre-persons. Physics and banks have addicted the United Nations, UNESCO and the universities to the "société anonyme" (French form of corporation).

My friend cast away his strongest weapon when he refused to call on the name of his noble predecessor. Using this name he could have created a realm in which one was not just a Peruvian or Yankee. But one has to admit to the existence of this realm by calling upon personal names; otherwise this realm will not exist. One can step into the light of history only by bearing one's first and last name. Bearing your name stops one from just belonging to a class or social stratum and disappearing into it. That's when a person rises above his social class costume; that's when his face shows above his clothing.

Whether we are generals, communists, pacifists, Chinese, or Americans, our own name can lift us above the generic ring in the nose, to a height at which we can look each other in the eyes and conclude peace. The spirit of peace and the peace of the spirit require that we think, act and decide by overcoming our race, class and our self-interest.

Technology, economy, psychology and sociology all gauge themselves by fine clothes, means of expression, circumstances and relationships. So they imitate the natural sciences. "Aid to developing countries" is their creature. And so they are afraid of men with first and last names, unless such people are their own staff members. Our history, however, has to be carried on by people who have names. Or it will end with the destruction of our race by statistics.

The modern devils are all schools of thought which deprive us of our names. We individuals are ridiculed by these madmen, because we would rather put up with a bad reputation than sink to the level of types, concepts, and numbers with which one does not speak. The Bolsheviks shut Tarsis in an insane asylum. But he had the courage to tell these so-called psychiatrists, "in my eyes you are simply policemen." Aid to developing countries is threatened with the same fate as Tarsis, that under the pressure of the spirit of the times people will prefer something nameless to men who follow the call to serve.

Applied to planetary service, no one is involved in the country where he is serving who does not love some inhabitant of the country, calling him by his first and last names. A country does not consist of its four million inhabitants, or of its sixty-thousand square kilometers, but is a place where I have lived so closely bound together with a name that I either love or hate, that we have moved in one rhythm of the days, of the joys and of the sorrows. Life means rhythm. Sociology however, is the lack of rhythm, the destruction of rhythm, because it wants to make us independent. But to live means to want to remain attached, simply to remain attached. What should one say to the pattern of life fostered by Bonn or Washington which never mentions attachment? The simplest way to prove such attachment, or dependence, is to use a beloved or dreaded name.

Life in service has to be life in service of a name, or the service remains lifeless. Nowadays people misuse the scientific method and think defining things in the abstract more important than calling men by their names. This threatens to limit service done for the planet to actions which can be comprehended by the antediluvian methods of worldly science and its empty generalities. "World" and "worlds" are nameless. The word "world" has intruded into the realm of speaking and listening beings in order to describe the nameless and comprehensible part of creation. God created heaven and earth, not the world. And only sons and daughters of the Earth who stop being children of the world are fit for planetary service. We only breathe with life of the planet as long as we are spoken to by name, and speak to others by name. When we do that we add to the orbit of our planet something which it needs from us in order to become a planet truly. As long as we do not listen and speak as inhabitants of our planet should, we hinder the planet Earth from coming into its own. The bureaucrats of abstraction, the theologians, the philosophers, the sociologists, the psychologists and the philologists are turning us into tempters, into pirates. But the reader already knows that we are using the word "pirate" in the opposite sense from the Biblican word "tempter." Just wait long enough and the lies of society change the clearest words into their opposites. After one thousand nine hundred and sixty-five years, it is laid on us to embody pirates.

The power to draw the dividing line between nameless worldly thinking and face-to-face planetary conversation will decide the fate of planetary service. Policemen who have lost their heads will always shoot, and party passions will always flare up. The two good boys in Venezuela and the four Yankees who were unsuccessful in Peru were not all unavoidable sacrifices. The two victims of the police certainly were unavoidable, because nothing will help against stupidity and fear. But we cannot call the four Americans an unavoidable loss.

What would it have been like if these four had grown close to four Peruvians before their service started? Duets, quartets and octets are organic preliminary stages of every step out into strange surroundings. When you have become one with brothers of an opposite group, even the most stupid spectators will be unable to reduce the group to its original components (Americans, German, Chileans, Persians). It is true that the young men and women are often too young to comprehend this preparatory stage in building groups. But this is where old age can and should jump in to require an intermediate step of them which would protect them from being classified cheaply as mere products of their particular background. In the Andean province of Peru, my American informant should never have permitted himself to call the creator of a university "a Peruvian." He also should not have appeared there as mere North American without a Peruvian comrade.

Such a chemistry of the soul cannot guarantee success either, of course. But we have tried it out, for example between the unemployed and academicians, Germans and Czechs, Jews and Christians and others. At least letting people know one another and building groups of friends beforehand, shows that we respect the size of the task. It does not leave everything to the stupid coincidence of an idiotic newspaper article. The first requirement for this is my calling my informant from Huamanga by his name. That is why I entitled this chapter "David Scott Palmer."

9. A Pirate's Esperanto

Peace is always something explicit. It has to be concluded in an area where no common language existed before. So this last chapter is going to concern itself with the coming language of the pirate's peaceful country. Our Esperanto won't be seeking words for bread and wine and money. It must proclaim a calendar of our tears and laughter common to all of us. People who keep peace among themselves share joy and sorrow. If they don't weep and joke together they certainly aren't at peace with one another. Pirates' Esperanto must fill the gaps left by the silencing of each group destroyed by technological progress. What we call urbanization today is the death of neighborhoods that lived by relying on one another totally. They are being replaced by a technologically-ruled world in which everything is preplanned.

In life, all laws sin against the fabric of shared time, the laws of the state as well as those of technological progress. In our national prisons there are convicts with sentences ranging from three days to life; they are put in a separate time. Our technological masterpieces fly to the other side of the moon; but neither in prison nor on the other side of the moon can we enter a life that would mold us together. The modern misuses of time, prison and technology, originated about the same time. The system of punishments using time evolved in Holland around 1700, and it was also in Holland that Descartes abstracted technological time from real time. Cartesuis was perplexed by our real time and he knew so little about it that he believed that God was creating time anew, every instant, without continuity between what had gone before or what would come after. By pushing buttons, technicians try to conquer the spatial part of life, for technological progress expands space although it shortens time. It is faster to turn on an electric light than to light an oil lamp. Every technological step expands space, shortens time, and destroys a familiar living group. This human group must be replaced explicitly.

A list of commandments for technology in the pre-industrial parts of the Earth arises from this law of progress of which everyone is aware instinctively, but which nobody consciously thinks about. Take a look at it: every step of technical progress destroys a familiar living group, with the result that human bonds of love have to be regrouped. That means permanent revolution. But why aren't people aware of this? Why have they conserved the costumes of the guards at the Tower of London? Why do our ministers preach every Sunday to the shards of old groups which they imagine to be church communities despite the fact that the groups have been long since destroyed by technology? Why don't they preach to bring together the new groups which need to be organized because electricity, etc. now exist?

The state started to murder our life-times even before technology did. Our penal system treats time as if it could be hacked into centimeters, into time-centimeters. That is why a high court can slice my life up like a loaf of bread when sentencing me to prison: three months, five years, or life. The penal system is based on the lie that six months are the same as twice three months, ten years are half of twenty, and so on forever gloriously. That is utter nonsense. I would be deadened after 10 years in prison. In keeping with their false "centimeter view" of time, the people of the 17th century invented the word "state" for an eternal institution with basic laws. They made laws for eternity. I have furnished proof of that at length in my book "About Industrial Law" ("Vom Industrierecht").

It is important for us to know that the statesmen prescribed devilish measured time for their beloved subjects long before the engineers did. All of your own experiences with time make a mockery of the penal code's hashed-up time. You experience time in the opposite way from space. If a person is damned to live his life by a stopwatch, his life-times will die. The madness of 1690, of arranging prison sentences in measured units of time like milestones, laid hold of philosophers and statesmen and accustomed them to the sight of the factory system, the logical extension of their own mistreatment of their subjects. Descartes, Spinoza, and Pufendorf issued a carte blanche for the future Taylor system by measuring the time of the children of men like distances on Earth. That contradicts all experience.

So every Indian, or Chinaman, or Papuan, or Watusi is head-and-shoulders above every prime minister and factory director in the industrial nations. For the Watusi still know that living man experiences his times in the opposite way from his spaces. And so he can laugh at us, kill us, cheat us, depending on what he feels like, but he would only be able to believe us after he had foresworn his innermost experiences, after he had become as uprooted as our whiz-kids in art and industry.

For spaces are experienced in the opposite way from times by generous and free-willed men. Babies in cradles stretch their arms toward an undivided universe. Boundless, that is the impression which space makes in the young world-conqueror. We arrive at our borders in space later. "The youth sails the oceans with a thousand masts, the veteran gratefully drifts into harbor keel upwards." But how do we enter into times? The opposite way. A baby knows nothing about an ocean of time; time does not stretch out ahead of him or behind him. For he thinks that he belongs to the moment. Pindar's vision of man as a fly with a one-day life span is everyone's first experience. Each small increase in time enters the life of a child as a gift from his community. His birthdays are given him by the love of his family. Birthdays, Sundays, holidays, the change from school to vacation; all times are social gifts to me. Expressed formally: I myself only experience time as a second. I can experience it as eternity only thanks to the whole human race.

Thus every step of technological progress destroys or at least threatens a step of human progress, because it shortens the periods of time which mankind hitherto constructed and expanded with unending faith. Factory time threatens the entire historical order of Upper Bavaria or Vietnam. A farmer in Bavaria may sigh, "It doesn't matter, I've gotten used to it". But what should one say when Point 9 of an aid program reads: "We must export human beings." Will no one arrest this impudent fellow? Human beings may well export: automobiles, coffee, or chicken legs. But even when the best subjects of the German princes emigrated to America in 1849, the prince didn't dare to boast, "We are exporting human beings." Even the Hessians who helped the British in America were considered auxiliary troops. Slaves were exported. What madness has been produced by the affluent society, that its trained economists presume to include the sentence: "We must export human beings," in their program?

As the sniveling statisticians protect themselves by declaring every laughing, crying, calling, or singing man unscientific, I can only leave the reader to decide. Should he be willing to burn their sentence "We must export human beings," as Luther did the papal bull at the Elster gate, then he will agree with me that we have to defend ourselves against these spiritual slave-traders. Our means of protecting the economy would be to deport these criminals, these economic experts who have unmasked themselves as spiritual slave-traders, to Devil's Island. To deport is still more human than to export, for that at least, springs from our indignation.

But in the academic world anyone who can speak even more indifferently, even more unconcernedly, even more voicelessly than his predecessor, is always considered even more correct. The more unconcerned, the more superior. I, certainly, cannot write the sentence, "We must export human beings" without crying out inside. Therefore I am ridiculous in the eyes of those who manage the economy nowadays. They have established the glorious rule: he who cries out is wrong. Thanks to this rule they are right. They will call forth a Sicilian Vesper in the world, which will cruelly murder the men they are exporting as well as others: Alas! What have they done? Industry profited from Auschwitz.

But you, reader, must understand one thing from this: you should not expect agreement from these "economic leaders." Don't let yourself be sent by them, for they won't send you unless you can be exported. You can be exported, but then you would not be an envoy. Were you exported, not dispatched, you would belong on the side of technological progress and the times of prison sentences. You would serve five years in India, four years on the

Zambesi just as you would in the peace-time army. But that would be useless time, because it is reckoned in terms of expenses. The true time, your time, should not cost; it should yield. I would be sorry if you were to throw this little book away now, impatiently. Because only now might I be able to put piracy into the framework where it belongs, into the framework of peace brought forth by the time of our future history.

I was screaming so rudely just now to stop you from crashing down into the technological age. Technology may concern itself with button-pushing, with speeding up the slow times of earlier ages. We, however, have to replace the groups destroyed by technology, and to do that we must voluntarily slow down. A man from the industrialized world must separate himself visibly from the technology that he brings by employing time values in his life with the foreign people opposite from those of the machines he has brought. Machines speed things up. The pirates' peace arises from a voluntary patience, an unexpected and technically unnecessary slowing-down.

At this point, we should renew our connection with the ending of the chapter before last, with that lovely expression "one doesn't need to know more." The girl wrote that about her own activities. We must keep this willingness "not to know more" in mind. Assuming this willingness not to know more, I will arm you with more exact knowledge of the horrible time diseases which threaten every industrialized modern car driver. You, everyone, should want to know more about that, for *otherwise everything will fail.* Confusing exported articles with men setting forth will produce three disasters in regard to time in the volunteer who has not learned the ABC of time as well as the rest of the language of the foreign country. The exporter, the engineer, and the planner are missing vital powers of life these days. They have never heard that this lack makes us illegitimate, that it does away with the present. As illegitimate children of the world in a vanished present, even with the best intentions we can only cause disaster.

I am well aware that I could create the impression of being a practical man here at the end of the book, by offering the reader an address book of all the organizations active as peace service corps, as development experts, as reconciliation corps, as civilian services, as work camps, or as emergency aid groups. I could also append figures and maps. I would give the appearance of sitting in a large office with a filing system, of easily disposing of all inquiries, like a computer. Each reader would be helped immediately. But it is not like that. I cannot dispose of any inquiries, I can only answer questions and then only if I have been plagued by the same question myself. Inquiries are so different from questions. Once you have turned away from mass demonstrations, acts of parliament, and newspaper articles, toward the questions of your own life, I would like to equip you not with phrases but with the speech aspired to by all pirates of peace.

This language begins with the mysterious twins "war and peace." These two words did not exist in the Indo-European language. This fact alone is thoroughly puzzling. Apparently we men lived a long time without making a distinction between war and peace. That these names were once missing can be understood after the chapters of this book. Both our ancestors and this book distinguished between three steps which modern humanity does not recognize: the speechless stage of the pre-human world, the declaration of war by the estranged children of Adam, and the explicit conclusions of peace between the reconciled children of Adam, of Cain, of Abel, right up to us.

When we distinguish between these three steps, twilight dawn, morning strife, and evening conclusion of peace, we experience time in three entirely different ways.

The first is living for the moment. Laughing and crying, hunger and love, sleeping and being awake alternate their reign over us. Language calls them "urges" for a very good reason. When we are passive, our "urges" urge us through time, and throw us abruptly into uncertainty like water falling from ledge to ledge.

The warrior emerges from this impulsive life thanks to his higher purpose. There is the politician who pulls himself together to face higher goals only because of his enemies. War concentrates time, it creates an affinity between many moments, reaching forward and backward in time. For the duration of the war all urges are subordinated to the one goal of victory. So everyone serving the community is elevated to working on long-range assignments. Our will-power achieves its height in times of war, when fighting off an evil. It lets us accomplish miracles of professional skill. Necessity is the mother of invention. Statesman and technician are both fighting the foe outside, be it the enemies before the gates of Rome, like Hannibal, or famine before the borders of Russia in 1963. Once stated, the watchword "fight the enemy" clarifies all the individual steps toward the defeat of the enemy. During the war, the enemy seems to be someone who has nothing to say to me and who refuses to listen. War means breaking off relations. The declaration of war is the last word exchanged by the parties. From there on they no longer have the time in common. That fact is very important, because it has been entirely forgotten.

How is peace concluded? By re-establishing both a common time and a common calendar among enemies. Peace cannot be achieved by an armistice, for then a cold war just replaces the hot one. East and West are not supposed to share common sorrows and common joys. On the contrary, there are only triumphs over the East or over the West, depending where you are. Where days of sorrow and days of joy are not celebrated together, either twilight or war exist, but certainly not peace.

And so because the conclusion of peace produces a time shared in common which would not exist naturally in spite of very constricted space, no one in industry understands peace. Industry is like a parasite for which peace is a prerequisite. It exploits peace. With its technology it works towards a shortening of time. You won't find the thought "should it perhaps go slower?" in the thinking of natural scientists. How did Galileo and Planck first achieve a common time? Why do Madame Joliot Curie and Otto Hahn share the same joys and worry about the same accidents in their work? Dear reader, the technicians, the physicists, and the chemists understand themselves as little as Bismarck or Adenauer understand themselves, let alone their wives. You and I may not understand atomics, but we can understand what drives Mr. Einstein and Mr. Bismarck, and we can understand it better than can these gentlemen, for that is our profession as pirates of the peace. That actually brings out the first great secret of time.

Time without a calendar of the soul's sorrows and joys shared in common, is an abstract time, is time artificially reduced to measuring external things, like police, labs, walls, borders, and concepts: office hours, the working day, term of service are times like that. Our time, however, has nothing to do with Bismarck's or Einstein's time. If they want to partake of life they must come over to us, the volunteers of time. For we cultivate the true character of time which everyone shares. They too need this sort of time in order to lead fruitful lives but they have to borrow it from us. For real time is time shared by several generations. During a war, in a fight, or in the strife of politics and nature, the living generation does assert its own will. Every peace, however, will connect several generations in such a way that the will of any one generation will become unimportant or undefinable.

In the 20th century much has been said about communal will, plebecites have been organized, and the percentage of people voting has been raised. Ninety-eight percent voted in favor of Hitler in 1935, and in 1945 ninety-eight percent would have voted against him. Shortlived as I am I always will remain stupid. The chain of generations however, is wise, as Edmund Burke said. One generation after the other has to be won over explicitly and induced to conclude peace. It is no secret, nor is it incomprehensible either, that peace can be concluded only step by step. No peace can be concluded like an armistice. Telephone and radio make it possible to have the shooting stopped within seconds. The opposite is true of peace. Three generations have to be pulled and led into a peace before it has come about.

An engineer who proudly presents his patented turbine to the curious Cambodian does not understand peace, because peace has to level us. To level us? What sort of crazy words are those? We have no need of that, the technicians will say. But that is what peace rests on. Peace can only come to people whom it levels into the chain of generations. In war we remove our will from this chain for the sake of a contemporary goal, of a particular moment. After such an outburst for the sake of a passing purpose, peace seals the times together again. And common joys and sorrows become again more important than the passing goals of wartime.

So it means that peacetime always feeds and fills our consciousness with several lifetimes, while war does so with one lifetime. That is why peacetime can be produced neither by you alone nor by all of your contemporaries. Peace addresses itself to "distemporaries", to men whom we have not even bidden a good day hitherto. Every peace which is more than an armistice connects and unifies people who have belonged to different ages. TV-watching, telegraphing and flying mankind is famished for this kind of peace. In New Guinea there is an explicit peace in force between two tribes which was agreed upon 6000 years ago. In Central Europe we have an unsettled state of affairs, neither peace nor war. It exists because technology is in control, and technology uses up our common time. It consumes the times which have produced our holiday calendars and the years of our souls for millenia. People should not run to the time consumer to put in an order for a new, future time. But where should we go?

For the moment we must come to ourselves. The words "come to ourselves" show that we are turning away from technology. They do not mean that we are growing silent, on the contrary, coming to oneself makes people most eloquent. The pearls of speech are the fruit of coming to yourself. But what does that foolish "coming to yourself" mean? Well, becoming indifferent to your status, your reputation, your appearance, your rank, your honor, or your rights. It shouldn't bother us, if someone calls us a Jew, or a Red, or an Aristocrat, or "a bit Catholic." Actually we can only come to ourselves if we don't care whether the world considers us legitimate or illegitimate, decent or indecent, noble or plebeian. And what else would show that we had come to ourselves! These are precisely the judgments that we want to sort out anew.

So let's take the boldest step. Is the whole era to which we naively lay claim *illegitimate*? What claim do we have to the time that I should be allowed to call it "mine"? There will be less strife, if the reader and I together would come to ourselves and for once call our own time "illegitimate", instead of reserving that label for the baby Jesus. Then our successors might again become children of honorable people; that is legitimate.

In our history the times of peace and the times of war were

kept separated. We have trained first rate soldiers and first rate craftsmen. The craftsman has an unending amount of time, the soldier has to be fast. The world situation today tests the national characters in a completely new way; it demands that we now overcome each day half as a soldier, half as a craftsman. Both ways of experiencing time, having no time, and having an eternity at one's disposal, must be alternately available to us in the factory, in politics, at home, in commerce.

We need both the planner and the unplanned man every single moment, or we will be ruined. The old divisions to which we are accustomed between long peacetimes and short periods of war, are yielding to a continuous double-time in which both types of time are welded together.

Life has fallen into deadening space of mere dead things where this welding together of both ways of experiencing time is missing. The present has then vanished and all our lovely plans and techniques are worth nothing. The planetary service fights this nothingness. Mistreated time becomes real time again when, against all expectations, we take our time. Repetition holds a great secret for the quality of real time, and anyone who takes time, takes time inside him- or herself, can receive Him Who is, Who was, and Who will be. The physicist who derives the present and then the future from the past is not the lord of time; but only a plumber. The word of the real Lord of time may strike our hearts in such a way that we can see both the future and the past through His all-present eyes. He tells us what must finally remain forever past, and what has to come our way without fail. Where His present goes before what we call "future" and "past" all of time is no longer in our mishandling hands, it is in God's hands.

Speech has long since expressed that. For the emergency has been obvious for a long time. It is only that our age of acceleration has chosen to ignore the other side of the coin: the need to have time galore. Let's approach this lost wisdom of speech slowly.

Things always go fast when we plan beforehand how to shorten the work load. Every book on planning tells the reader to start with the desired result and to work backwards, dividing the process into small steps. But after a new step in technology we should always go slower, or the sheer excess speed will leave no time for the next living group to gather itself together. If these men cannot gather themselves together, the result for which all hope, will not come to be. We cannot be forced to love. So the most impractical but most necessary gift that technological man can give the Lord of time and eternity, is the gift of free time in which someone can learn to like him. The planner must create a staff and that takes time.

If he could not also slow down time, man in shortening time would put himself in the place of God; he would no longer listen to the words of Bach's cantata Nr. 106. "God's time, God's time is the very best time." I well know that this sounds ridiculous, as the crafty stop-watch people reserve Bach's music for concerts. But cantata 106 does express a fundamental fact of our existence. The way we calculate time reflects only one half of our experience with it, that half of youthful hands ready for battle, and of clever minds proud of their concepts. The attack of hands and the concepts of minds are aggressive. They lead our hands and minds in the battle for existence. But our race will die out if it allows itself to be governed by hands and minds. For our word is true only when spoken from the heart for the whole race and when coming from our mouth, it reaches the ear of a partner. Reaching somebody is the opposite of attacking him. Expressing is the opposite of comprehending. A "good morning", or an "I forgive you" has nothing to do with comprehension; it must be expressed; that is, it must be risked in order to be heard. A concept puts an object in front of me. Speech, however, enters me to change me. And so God' time is the very best, because in its peace we can recognize that the parts of life which technology threatens to deliver into the hands and minds are in face unassailable and incomprehensible. God creates a time beyond the dead space of hands and concepts. We enter his time by suffering and loving, which happens the opposite way from the attack and the concepts which we meet in the aggressive individual. The whole genus man comes into its full play in peace. There is no neutral "individual". In the peace of mankind, no one can be an object to his brother because he becomes part of him.

As children we played a game called heaven and hell. Half a

piece of paper was colored black, the other half remained white, and after it had been folded skillfully we could alternate between heaven and hell. Each of us alternates that way between peace and war, between arming and disarming, between his concepts, and his personal relationships. When eating or giving orders your head-man and your stomach-man come into their own. Woe to them, if they do not take things in hand energetically. To do that, the one must comprehend and the other must attack. To beget and to create, however, we must renounce the whole armory of war. Love disarms; suffering overpowers. God's passion is the high tide of his creative power. Atheism is fine for experts, chemists and for Mr. Darwin who thought up the struggle for survival. But atheism would be funny only in someone who gratefully trusts in his greatest inspiration. For he is trusting God.

But far more is at stake. Faith that God's time can enter into you is the only justification for everyone of our rights. At the beginning of world history we had no rights. Fathers could kill their newly-born children. I can call this paper and pen my own while trying to equip the next pirate of peace only because the right to do so was granted me one fine hour although it was hitherto unheard of. Think of it, I can even criticize the authorities! Yes, nowadays a Pope can even criticize his own official position thanks to the rights of man from the French Revolution; for he too has become more than a Pope. Nothing which you naturally claim as a "natural" right, - your right to vote; your ability to swear an oath; your duty to serve as a lay judge, a juror, a selectman, or as a soldier; your freedom of speech, and above all, your right to choose your own profession and partner in marriage - was ours "in nature". In the Sistine Chapel, Michelangelo painted the path of these freedoms from God into men. Now, however, they are ours by law. In other words we should not believe that the current legal status of rights is based on nature. They can be lost again, and it can happen in the twinkling of an eye, as it did under Hitler. But that is exactly what half of your contemporaries who will read this will not admit to themselves, honored reader. They believe that they are in a state of nature, while possessing all of their rights. And so they will loose them all again at the next chance. And no one will give them back.

So let us remember: every basic right of free men was won in the course of an everchanging life. We did not possess it before then. Our fathers won it for us. So we can only claim to possess this legacy if we risk it and increase it while producing the next legacy ourselves. The battle which the pirates of our Peace Service Corps are about to join will again be fought over a basic right. It is my right and your right to be able to act in our own names, free of passports, states, or governments, despite the 140 states of the UN, despite the division of even the air space, and despite sovereign governments.

So a Pope shall help us to set up the roadsigns to the third millenium which we will not reach without a civilian form of service to set against the armies and the wars. The unknown soldier of peace and the name of John XXIII are graven on the same tablets of history. Even a Pope was confronted with the question of how the next necessary step that is lacking would be taken, how creation would continue. Can more than just one single lonely person take it? Mankind has been grappling with that ever since the Second Isaiah. The word "create" appears fourteen times in the Second Isaiah which is very short. Jesus drew from the Second Isaiah to complete the creation of mankind, of us, within himself. Thus we might be able to know the new man. And we must give expression to him in the Pirate's Esperanto, or we won't achieve piracy. The third millenium will not come to pass if we are administered by technicians only.

Pope John XXIII gave us this insight for our future when he was Cardinal Roncalli. He was papal nuncio in France in 1945. At this time an extraordinary enterprise was being talked about in Paris. The Scottish doctor, John W. Thompson, together with some other doctors had started an "anti-concentration camp" in front of the gates of old Lutetia. The doctors were hoping to heal the victims of the terrible "esprit concentrationnaire" in the camp, using the counter-poison method. The rules of their camp allowed everything forbidden by the Nazis. The residents were not inmates; they were tenants. They were allowed to come and go as they pleased. They could withdraw money from the camp-treasury or deposit money. They could work and sleep as they pleased. Cardinal Roncalli was given a tour by Dr. Thompson and found everything highly successful. At the end of the tour he sighed a little and turned to the doctor saying: "Isn't it remarkable that in our age the extraordinary things can only be done by lay-men?" This sentence will make it clear to the reader why Cardinal Roncalli then became the most extraordinary Pope. But the sentence also articulates the peace corps' secret for which we have been looking.

The technological world anticipates all functions by planning, and every expert is pre-planned and pre-programmed. He must accomplish something, something which has been known beforehand. With this he enters the world of space; space makes him visible. Since what is visible about us is prone to death, the spaces in which the experts act are not only dead but deadening. Only pure time is free of death, and a man can take part in it only by forgetting space and overcoming fear of death.

So a Pope ranked the unpredictable person - in his language the lay person - above the clergy. As early as 1926 I pointed out in a pamphlet "Religio Depopulata" that the clergy were losing their power. The Islamic warriors are losing their power these days, and they represent the clergy there. Truly important things in the Islamic countries and areas can only be done by the Mohammedan women now. Only non-professors are thinking and teaching important things in professional Germany today. Only non-gentlemen can start something which will catch on in England. In France, the provinces - Toulouse, Marseille, Le Havre, - which are the laymen the Pope spoke of - suddenly begin to be worth something, as opposed to Paris, the queen so far. Only a pirate, an unpredictable person, can save us from time-foreshortening, space-expanding, group-destroying technology. The former leading groups no longer contain the avant garde. The outsiders are the avant garde.

The good Samaritan, of course, is the original "layman", for a layman here means a person from whom things are least expected. It was natural for a Jew from Herod's kingdom to believe that all Samaritans would hate him. He himself certainly hated all Samaritans; the Samaritan whom we later called the "good" revealed himself as such in the course of the story. What the Pope discovered during his visit in the camp, and what the reader of this book may be discovering, is that the values "clergy" and "lay person", "decent man" and "pirate", "technician" and "lover" are reversing themselves.

The reader may have a nearly insurmountable difficulty which Cardinal Roncalli did not have. This son of a peasant from Bergamo had the gift of looking into the calyx of the simplest blossom as if he were seeing it for the first time, as Jacob Grimm said of Bettina von Arnim. This power alone, to see things afresh, makes it possible to bring forth the next historical moment. For the comprehensible part of our being, administrative authorities are indispensable: but incomprehensible man cannot help but make them impotent. They are not meant for this side of us, they have no responsibility there.

An American citizen is breaking the law these days if he takes a private trip to Cuba - 90 miles away. If the officials had their way, we would be allowed to follow only their directives. Even in the United States officialdom is shoving the proper conduct of affairs to one side. It was not always like that. I can remember a story from my youth, a dispatch sent by General von Einem to the Kaiser during a miners' strike in the Ruhr district, after the Kaiser had demanded a report on the situation. The general telegraphed: "Everything quiet but the officials". And we should obey only the experts? No, "war is too important to leave it to the generals".

So this is the sword of Damocles hanging over our future: not a single one of the administrative bodies of either the worldly states or the churches is responsible for the planet. They have sworn their official oath to one part of the world. It is incumbent on the lay person within us solely to step out of the jurisdiction of the individual state or church office, it is incumbent on us whether we are doctors or judges, it is incumbent upon the Pope himself, upon the professor as well as a minister of state. The most impressive precedent that I am aware of occurred during the Presidency of Theodore Roosevelt. The Russians and the Japanese waged war from 1904 to 1905, both sides suffering greatly. Roosevelt was asked to mediate, and most of my readers will know that the peace of 1905 was indeed concluded in the American harbor of Portsmouth. But listen! Roosevelt had merit which is never mentioned in the schoolbooks. He wrote to a friend in St. Petersburg, the American ambassador Lengerke-Meyer: "It would be in the interest of the United States if Russia and Japan were to continue weakening one another, but as President of the United States of America, I don't have the right to act only in the interests of the United States. I have to serve the higher interests of peace on Earth." The superclever textbooks from the Department of Education unfortunately leave out such expressions of laymanship even in heads of state. The German Kaiser William II can just barely be counted in the company of Theodore Roosevelt. William II wanted to stand up for Captain Alfred Dreyfuss of whose innocence he was well aware, and formally present his case to the French government. His chancellor, the unspeakable Bulow, stopped him, saying that it was in the German interest to let the French tear one another to pieces. Hearing that, the Kaiser did not act as unselfishly as Roosevelt did, but remained silent.

The super-clever always claim their fractional interest as God's whole. That whole is shattered by wrong doing. The witnesses to an unrepented crime become schizophrenic. The spirit of a single person breaks when crimes against the body of the whole are left unheeded. That is why justice is in fact undivisible although it seems divisable to an unjust man. Bulow's dream that French misfortune could become German luck is a nightmare, and made impossible by the common origin of both nations in Christendom. "God himself is law" wrote Eike von Repgow (1209-1232), in order that we, like white blood cells, dash to the damaged spot and reconstruct Him with one voice as His jurymen. Herr von Bulow was not in a hurry to expiate a terrible misfortune. The Kaiser was, for he was a living soul, and suffered from injustice along with the person to whom it was done. A person who does not suffer from an injustice done to his neighbor as if it were done to him should receive no help when wrong is done to him.

When she was a girl in 1916 a friend of ours who later became an architect and the mother of six children, read about the Turkish outrages against Armenians. Defenseless men were being mowed down, which infuriated our friend. America was not at war yet, and she went to Washington to work out some way of helping the victims. When she finally reached the only official who was even vaguely responsible at the State Department, he told her that help was out of the question, as anyone waiting to enter the United States had to carry a passport made out in his name, while the Armenians were a nameless bunch. "Then we'll send them a collective visa." "There's no such thing." "Yes, but the Armenians have to be saved." He shrugged his shoulders. When our friend remained seated, the official got nervous. "My office is closing". "Fine, I'll remain sitting here, until you come back tomorrow." "Miss, be reasonable. How do you want to force me to break the law?" "Oh, that's not hard. When you die and come before the gates of heaven, I'll be there. When you ask St. Peter to let you in, I'll say to him: Don't, he wouldn't let the Armenians in." A collective visa for the Armenians was invented, it was issued, and I have met an Armenian myself who is a US citizen now, thanks to this collective visa. In those days, the officials still recognized that doing the proper things ranked above officiality.

If we listen at such a unique moment, the creator of Heaven and Earth can enter us anew. Each of us is required to let God's law put our heart under such pressure that our lips thereupon speak to our neighbor about it. Only by again saying - each of us - the right word to our neighbor from the bottom of our heart, can broken justice be spoken anew, and the future of our planet be put at ease.

Of course the bureaucratic whiz kids reduce the real world to a state of affairs for which they already carry responsibility. These officials come from Nineveh and Assur; from Babylon and Luxor; from all the semitic, hamitic, or japhetic areas of the devil, and with this rubbish of Osiris, Marduk, Wotan, and Fitzliputzli as the cosmic and comic source of their decrees they teach: "Long live injustice! An injustice done to my neighbor will protect me. If I sacrifice him, I'll be all right." This statement robs a courageous soul of his home on God's planet and in our time. So the statement which the future Pope made in Paris becomes more important with each passing year of our bureaucratization.

The unpredictable, referred to in the Pope's statement, offers the youth of the world a more important dogma than his ecumenical council can. An ecumenical council just liquidates a past which has become all too human. The amazing statement which Pope John XXIII made when a cardinal, on the other hand, opens the floodgates of an immeasurable future.

For this reason, the old *doctores* of the Church named the Creator "actus purissimus," the purest act, they called Him "time without space" so to speak. Only we, His creatures, fall down into space insofar as we will die. Pope John only re-formulated this sentence. For a "layman" has no place in the world, no position, no contract; he does not belong to the personnel of any factory. Nothing he will achieve is known beforehand. He does not execute any plans. He is a pirate. A layman in this sense is the person who cannot be made part of a plan, the unforeseen and unpredictable man. The further technological plans progress, the more we will have to make use of the unplannable, free time of the pirates of peace. More surprise and more drawing on your reserves should out-balance increased progress and increased planning. Tragically, since 1917, the technicians claim to be able to heal us with their progress, and this has accelerated the war of everyone against everyone else. They lost sight of time as they concentrated on space and the machines which control it. "Free time" contractually guaranteed to man working in industry is a caricature of time. For it is only scraps of time left over from work, a remainder of the day in which you leave work to go home. It is the sort of time available to slaves. Only days of celebration are free time. Men who want to establish peace together in the stream of eternal life from A to Z- the Bible calls it from alpha to omega - have to offer one another true, pretechnical time, not the time of watches. Such men stop the time of watches. They establish the calendar of holidays anew.

We would be capable of celebrating holidays were we not surrounded by the noise of television and radio. But they crash around us nowadays, making everything happen simultaneously with an unheard of pressure of impact. They snatch all of our senses away from the slow purpose of our life. Advertisements are the most obvious example of such pressure and impact, but advertising is only the most obvious case. All electronic media produce instant programs. Therefore they take clock time terribly seriously, and they carry us off into all shortcuts that they can reach. But woe to us if we are carried away by them. Then there will be no end to the rush of measured time outside ourselves. The layman's time to which the future Pope referred, the time of the good Samaritan, the time of the pirates is time measured inside oneself, and therefore an immeasurable time. At Easter, at Christmas, and at Pentecost we have that sort of time galore. Future cannot then be told from past, for the reason, forgotten today, that there the present of the celebration unfolds from both, from the end and from the beginning thanks to our inner peace. The celebration is raised above the calendar of everyday life because all times participate in it, and no ostentation can replace this participation. The rank of our holidays, whether mere noise or true celebrations, is determined by the extent to which all times participate in them when they occur. Everywhere the molecule of peace must be able to attract and hold the willing human atoms who have completely different religions yet are moving so close to each other these days. This is what the service on the planet must achieve.

Speech has long since expressed this secret. It is just that the militant number-magic of technology has outshouted quieter speech. The pirate of peace will have to salve the planet with a drop of the oil with which the prophet anointed the king of peace, and with which all pious kings have been anointed ever since. The pirate, layman, or Samaritan who remains unrecognizable is the uncrowned king of today. But even if uncrowned he must not remain un-anointed. His calendar unofficially continues the calendar of the saints. Don't laugh at me. I am not getting unctious just because I dare speak openly of anointing the princes of peace. The word "anoint" only means that someone must be freshly prepared for something which awaits him in the future. When a bride leaves her parents' house, her father used to put the same burden upon her "The young bride leaves her parents' house: go, love, and suffer." No one, least of all the bride, can guess what will be demanded of her on the morrow. Joy or sorrow? Always the unexpected. In this, every soul is like a bride. The unrecognizable kings of the third millenium are indeed all uncrowned, but like the shepherd boy, the son of Jesse, all may know themselves anointed by the prophesy that the world is waiting for them in order to become a planet. World government, what a scourge of God! The planetary peace, how light a yoke.

Speech asked us audibly long ago to stretch the inner time span between equipping ourselves for peace and its explicit enactment. Why? Well, our technology puts its achievement before us. Here, customer, is your refrigerator; here is where the trains run, here is the radio. Now use them. This is what it gives me to understand. It makes us into masters of the world. Peace does none of that. It cannot be used. For it has need of us. It dominates us. Without anointment, without our exertion, and without our inner dedication we will never be at peace within the peace which we have achieved. For then we have not recognized its domination. The dividing line between war and peace, between technology and holidays runs just there. The wicked world can force us into war. But we can be forced neither into peace nor into life. The soul has to enter peace explicitly and voluntarily. "At peace" is an amazing phrase. A contented man is "at peace". The words "at peace" express an understanding that peace lies beyond us and not under our control, an understanding better than that of the methodical, practical man. Peace goes beyond you and me. But we move towards it. So speech tells us that peace is an eschatological word! It is only promised to a person who yearns for it. No planner can comprehend that. But it is true all the same. Peace cannot come to exist without first being longed for. Otherwise it is not being at peace, but vegetating. Peace merely willed by you is cold war. Technological foreshortening of time can be offset and overcome only by a service to the planet which is being patiently "at peace", and proceeding towards peace. We have the authority to sacrifice measurable moments, independent positions, our office: in one word self-willed time, for the future peace. Our race owes every one of its days of peace to this authority alone.

And therefore, dear reader, I did not want to furnish you with technical, official knowledge about the future planetary service. You can easily find such technical advisers. Before they come to visit you, I had to hand on to you something which since about 1910 has entered into a race with the longings of the national warriors. Honor to their victories. The more so as we ourselves can hardly boast about the victories which we have already won. But warriors have been ennobled not by being cocksure and boasting, but by forgetting themselves. When this happens a new language is born between the one person at home somewhere and the one person trying to help him be at peace there.

Index

Names with "de" or "von" are listed under the letter of the name that follows.

Eugen Rosenstock-Huessy (1888-1973)

Now known as a social philosopher, Eugen Rosenstock-Huessy worked and taught in a wide range of fields. He brought the full range of western tradition to bear on the challenges posed by the two world wars and the coming centuries.

He was the son of a German banker, and became a professor of law at Leipzig University when he was 24 years old. In 1914, he entered the German army and was an officer near Verdun during World War I. At that time, he and his friend Franz Rosenzweig conducted an extensive correspondence on Judaism and Christianity. Their letters, first published in the 1920s, are now considered a classic contemporary exchange between Christian and Jew.

The same year in which he entered the army, Eugen Rosenstock married Margrit Huessy, who shared in his endeavors throughout her life. After the war, Rosenstock-Huessy worked at the Daimler-Benz factory, founding and editing the first magazine specifically for factory workers in Germany. In 1921, he founded the Academy of Labor in Frankfurt. Two years later, he began teaching at the University of Breslau, where he taught the history of German law, as well as civil, commercial, and industrial law. During that time, he began to articulate his proposed method–which he based on speech–for the social sciences, by publishing *Soziologie*. In 1931, he published *Die europäischen Revolutionen,* a book establishing him as a scholar and writer in Europe. He produced a different version of it for English-speaking readers with the 1938 publication of *Out of Revolution.*

In the period between the wars, Rosenstock-Huessy was active in establishing work service camps in Germany, bringing together students, workers, and farmers. Several of these early work service participants later became an important part of the German resistance to Hitler.

Vowing never to work a day under Hitler, Rosenstock-Huessy left Germany and emigrated to the United States. He eventually accepted the chair of social philosophy at Dartmouth College, and taught there from 1935 until his retirement in 1957. With the support of President Franklin Roosevelt, he built on his experience of creating work service camps in Germany and organized an experimental camp within the Civilian Conservation Corps. World War II interrupted the program before it had come to fruition.

Throughout his life, Rosenstock-Huessy continually wrote books, papers, and monographs, and gave talks and lectures to many generations of adults and college-age students throughout the United States and Europe. *The Christian Future* appeared in 1946, and an expanded 2-volume *Soziologie* was published between 1956 and 1958. His major work on the speech of mankind, *Die Sprache des Menschengeschlechts,* was published in 1963. He died at his home in Norwich, Vermont, in 1973, at age 84.

Bibliography

Books by Rosenstock-Huessy

available from Argo Books • RR 2, Box 366A • Jericho, Vermont 05465 • (802) 899-4883 phone and fax

The Christian Future, Harper Torchbooks, 1966, 248 pages, paperback.

Fruit of Lips, or Why Four Gospels? Pickwick Press, 1978, 143 pages, paperback.

I Am an Impure Thinker, Argo, 1969, 200 pages, hardcover.

Judaism despite Christianity, the "Letters on Christianity and Judaism" between Eugen Rosenstock-Huessy and Franz Rosenweig, Schocken, 1971, 181 pages, paperback.

Life Lines: Quotations from the Works of Eugen Rosenstock-Huessy, Argo, 1988, 83 pages, paperback.

Magna Carta Latina, Pickwick, 1975, 296 pages, paperback.

Mutliformity of Man, Argo, 1973, 78 pages, paperback.

Origin of Speech, Argo, 1981, 160 pages, paperback.

Out of Revolution, Argo, 1969, 795 pages, paperback.

Planetary Service, Argo, 1978, 144 pages, paperback.

Practical Knowledge of the Soul, Argo, 1988, 66 pages, paperback.

Rosenstock-Huessy Papers, Vol. I. Argo, 1981, paperback.

Speech and Reality, Argo, 1969, 200 pages, paperback and hardcover.

Other titles relating to Rosenstock-Huessy

Camp William James, by Jack J. Preiss, Argo, 1978, 272 pages, paperback and hardcover.

A Complete Bibliography, compiled by Lise van der Molen, Edwin Mellen Press, 1989, 147 pages, paperback.

Eugen Rosenstock-Huessy, Studies in His Life and Thought, edited by M. Darrol Bryant and Hans R. Huessy, Edwin Mellen Press, 1986, 270 pages, hardcover.

Letters to the Third Millennium, by Clinton C. Gardner, Argo 1981, 272 pages, hardcover and paperback.

Speech and Society, The Christian Linguistic Social Philosophy of Eugene Rosenstock-Huessy, University Presses of Florida, 1987, 220 pages, hardcover.

Lectures by Rosenstock-Huessy
full transcriptions and audio cassettes from Argo Books

The Circulation of Thought-1954, Dartmouth College; 26 lectures on the sociological aspects of knowledge, covering the process of thinking in personal life, education, and science.

Comparative Religion-1954, Dartmouth College, 24 lectures discussing philosophy and religion, American religion, religion and timing, religion and speech, and religion as the power to undergo change.

Cross of Reality-1953, Dartmouth College, 24 lectures introducing Rosenstock-Huessy's sociology, and outlining his method of analyzing human social institutions.

From Karl Marx to the Peace Corps, 1954 and 1966, 5 talks ranging from technological change to work service, and addressing the failure of the liberal tradition to give social catastrophes a central position in social theory.

Greek Philosophy-1956, Dartmouth College, 26 lectures covering major Greek philosophers; and philosophy versus prophecy, versus religion, versus the arts and sciences, versus prayer, versus Christianity.

Hinge of Generations-1954, Dartmouth College; 14 lectures addressing history as a social process among generations.

Saint Augustine-by-the-Sea, 1962, Santa Monica, California; 6 talks covering language, speech, names, grammar, teaching, renaissances, experienced time and Christianity.

Universal History-1954, Dartmouth College; 25 lectures covering the social histories of tribes, astral empires, the Greeks, and the Jews, as well as historiography.

Universal History-1955, Dartmouth College; 4 lectures introducing history, with special emphasis on holidays, calendars, and the formal articulation of events.

Universal History-1956, Dartmouth College, 9 lectures covering the Holy Spirit, the arts and sciences, Judaism and Christianity as historical forces, the mind, nature, epochs, the Gospels, Church and State, generations, and holidays.